Exploring The BUILDING BLOCKS of SCIENCE

Book 3

LABORATORY NOTEBOOK

REBECCA W. KELLER, PhD

REAL SCIENCE 4 Kids

Illustrations: Janet Moneymaker

Exploring the Building Blocks of Science Book 3 Laboratory Notebook
ISBN 978-1-941181-02-7

Published by Gravitas Publications Inc.
www.realscience4kids.com
www.gravitaspublications.com

Contents

Experiment 1

A Day Without Science

Introduction

Imagine you are living a day without any of the technologies that science has helped us create. What would your day look like?

I. Think About It

❶ Do you grow your breakfast or do you get it from a store? Why?

____

____

____

____

❷ Do you make your clothes or do you buy them? Why?

____

____

____

____

❸ Do you ride a bike or do your parents drive a car to take you to school? Do you get to school a different way? Why?

____

____

____

____

❹ Do you use electric lights in your home to read by? Why or why not?

____

____

____

____

❺ Do you use a heater or a cooler to keep your house just the right temperature? Why or why not?

____

____

____

____

❻ Do you watch TV, listen to music, or text a friend on a cell phone? Why or why not?

____

____

____

____

II. Observe It

❶ Carry this *Laboratory Notebook* with you for a day and list all the things you use that have been created or invented with science.

Things Invented with Science & Used in a Day

❷ Describe what your life would be like if you didn't have any of these items to use.

Life Without Science

III. What Did You Discover?

❶ What types of toys or electronic gadgets did you use today? Was science involved? Why or why not?

---

❷ Did science make it easier for you to travel today? Why or why not?

---

❸ Was science involved in anything you used to see better during the day and the night? Why or why not?

---

❹ Would your life be much different if there were no scientific discoveries? Why or why not?

---

❺ What things did you use during the day that did not involve science?

IV. Why?

Today, most people use a wide variety of items that have been invented with the help of science. As knowledge of chemistry, biology, physics, geology, and astronomy has advanced, new discoveries and inventions have made our lives easier and have made it possible for us to do things we weren't able to before.

We now have cars and trucks, airplanes, and speedboats so we can travel faster and more comfortably. We have satellites that help us communicate and navigate and spaceships to explore other planets. Computers make it possible to collect and interpret data in ways that couldn't have been done before.

Even most of the food eaten today has been grown or modified using science. Clothing, housing, and even education have all been changed by applying scientific knowledge to problems that we wanted to solve or new things we wanted to create.

Each science discovery broadens the foundation for new discoveries to build on, allowing scientific knowledge and applications to proceed at a faster and faster pace.

V. Just For Fun

Imagine you live 2000 years in the future. Write or draw what you think the world might look like. What new things do we know and what new inventions have we created?

Experiment 2

Make It Mix!

Introduction

What happens when different liquids are mixed together? Try this experiment to find out.

I. Think About It

Look at each item in the first row of the chart below. If you think the item is "like oil," put a check mark in the row labeled "Like Oil." If you think it is "like water," put a check mark in the row labeled "Like Water."

	Water	Milk	Juice	Oil	Butter
Like Oil					
Like Water					

❶ What do you think will happen if you add water to milk?

____

____

____

____

❷ What do you think will happen if you add milk to juice?

____

____

____

____

❸ What do you think will happen if you add juice to oil?

❹ What do you think will happen if you add oil to water?

❺ What do you think will happen if you add oil to melted butter?

❻ What do you think will happen if you add soap to water and then add oil?

CHEMISTRY

II. Observe It

Mix two liquids together and record below whether they mix or don't mix. Use about 15 ml (1 Tbsp.) of each liquid when making the mixtures and use a separate cup for each mixture. Label each cup with the contents.

Results of Mixing Liquids

	Water	Milk	Juice	Oil	Butter
Water	X				
Milk	X	X			
Juice	X	X	X		
Oil	X	X	X	X	
Butter	X	X	X	X	X

Observe It With Soap

Take your mixtures from the first part of the experiment and add soap to each mixture. Note whether the addition of soap changes the way the liquids mix. Record your observations in the chart below.

Results of Adding Soap to Mixtures

	Water	**Milk**	**Juice**	**Oil**	**Butter**
Water	X				
Milk	X	X			
Juice	X	X	X		
Oil	X	X	X	X	
Butter	X	X	X	X	X

CHEMISTRY

III. What Did You Discover?

1. What happened when you added milk to juice? Did they mix? Was this what you expected?

____

____

2. What happened when you added water to juice? Did they mix? Was this what you expected?

____

____

3. What happened when you added oil to water? Did they mix? Was this what you expected?

____

____

4. What happened when you added soap to your mixtures? Did the soap, water, and juice mix better?

____

____

5. Did the water and oil mix better with soap?

____

6. Did the oil and butter mix better with soap?

____

IV. Why?

When you added milk to juice, or water to milk, you should have seen that they mix. *Mixing* is when two things blend together so well that you can no longer tell them apart. Milk is white and juice is colored. When they are added together, the mixture of the two is a new color, a cloudy color. Juice and milk, water and milk, and juice and water mix because they all are the same type of liquid.

What happens when you add oil to water, or oil to juice? Do they mix? You should have seen that when you added oil to water or oil to juice, they did not mix. No matter how much you try to get oil and water or oil and juice to mix, you will always be able to see little droplets of oil floating around, not mixing. Oil and water cannot be mixed because they are *different* types of liquids.

You should have seen that when you added oil to butter, the two mixed. Why do you think this happened? It happened because oil and melted butter are the *same* type of liquids.

The rule is: *Liquids of the same type mix and liquids of different types do not mix.*

What happens with soap? Soap is a little bit like water and a little bit like oil, so soap can make water and oil "mix" a little bit. This is why you use soap to wash oils off your hands!

V. Just For Fun

Think about different ways you might test liquids to find out whether they are "like water" or "like oil." Use one of your ideas to create an experiment to find out if the following liquids are "like water" or "like oil."

→ Soda

→ Coconut oil (or olive oil, canola oil, or other vegetable oil)

→ Orange juice

→ Mayonnaise

→ Another liquid of your choice

Record your observations in the chart on the next page, or make up your own chart. Name your experiment and write the name on the chart. Below the chart make notes about how you performed this experiment. Also record whether or not you think the experiment worked and why or why not.

Experiment 3

Make It Un-mix!

Introduction

Do this experiment to discover some ways to separate different kinds of mixtures.

I. Think About It

❶ If you had rocks and Legos mixed together, how would you un-mix them?

____

____

❷ If you had rocks and sand in a bag, how would you un-mix them?

____

____

❸ If you had sand and salt in a bag, how would you un-mix them?

____

____

❹ If you had salt and sugar in a bag, how would you un-mix them?

____

____

❺ If you had salt and sugar in water, how would you un-mix them?

____

____

CHEMISTRY

II. Observe It

❶ Take a handful of rocks and a handful of Legos and mix them together on the table. Now try to un-mix them. Draw or describe what you did.

❷ Take a handful of rocks and mix them with sand in a bag. Now un-mix the rocks and sand. Draw or describe what you did.

CHEMISTRY

❸ Take a handful of sand and a handful of salt and mix them in a bag. Now un-mix them. Draw or describe what you did.

❹ Place a few drops of several different colors of food coloring into a glass or paper cup that contains 120 ml (1/2 cup) of water. Think about ways you might un-mix the colors.

❺ Now try a method called *chromatography* which uses paper to un-mix the colors.

Take a piece of coffee filter paper and cut it into long strips. Place a pencil over the top of the glass or paper cup that contains the colored water mixture and tape one end of a paper strip to the middle of the pencil. The other end of the filter paper will be in the colored water.

It should look like this:

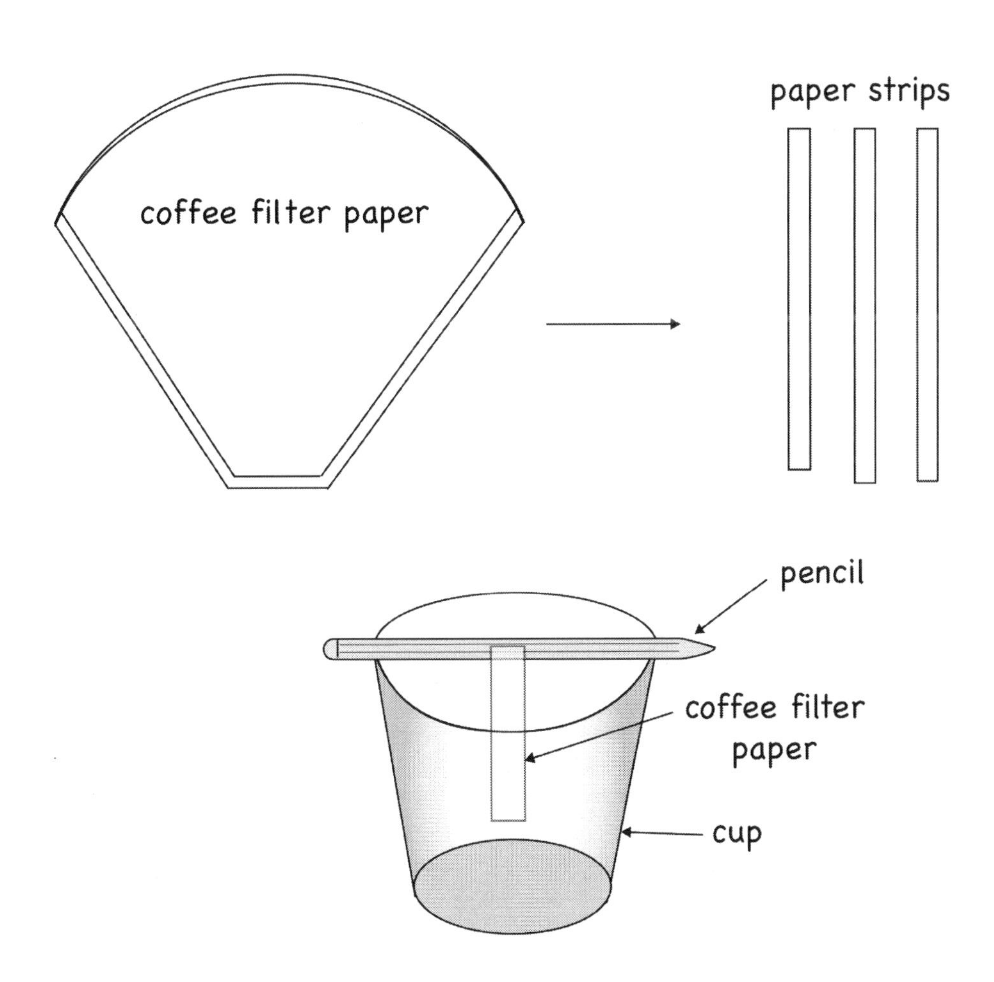

❻ Carefully observe what happens, and record your observations.

CHEMISTRY

Let the filter paper dry and tape it in the box.

7. Repeat the experiment with an "unknown." Have someone mix two colors together in water. See if you can tell, using chromatography, which colors are in the water.

8. Make an unknown for your teacher. Mix two colors together without telling your teacher what they are. Have your teacher use chromatography to find out which colors you put in the glass.

9. Record all of your results below.

Your Unknown

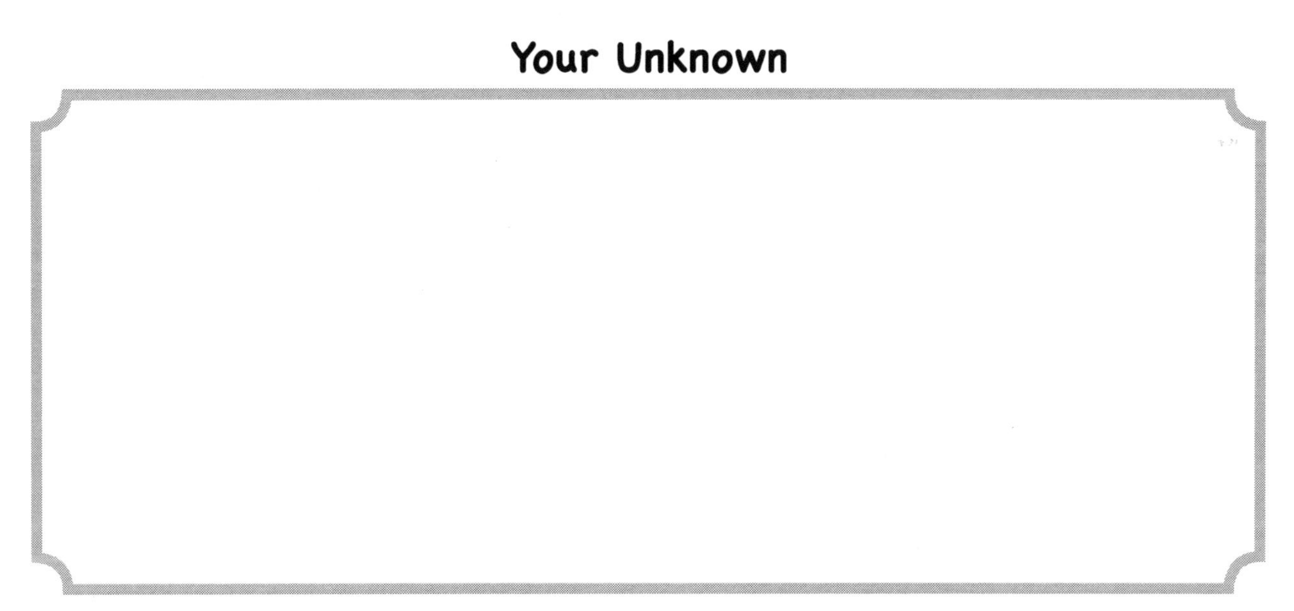

Teacher's Unknown

III. What Did You Discover?

CHEMISTRY

❶ How many different ways did you discover to un-mix things that were mixed? List them.

❷ Can you use your fingers to un-mix the rocks and Legos? Can you use water? Can you use paper? Why or why not?

❸ Can you use your fingers to un-mix the sand and the rocks? Can you use water? Can you use paper? Why or why not?

4. Can you use your fingers to un-mix the sand and the salt? Can you use water? Can you use paper? Why or why not?

5. Can you use your fingers to un-mix the colors in the water? Can you use water? Can you use paper? Why or why not?

IV. Why?

You found out in the last experiment that some items mix and some do not. In this experiment you discovered that sometimes you can un-mix items and sometimes you cannot. You also discovered that large objects, like rocks and Legos, are easier to un-mix than small items, like salt and sand. You discovered that items that look similar, like salt and sugar, are difficult to un-mix.

Why are some mixtures hard to separate and some easy? Larger objects, like rocks and Legos, are easier to un-mix than smaller items like sand and salt. Items that have very different properties, like sand and sugar, are easier to separate than items that are very similar, like salt and sugar. Also, very small objects that are hard to see are very difficult to separate. For example, molecules are very difficult to separate from other molecules.

You found out that when items are difficult to un-mix you can use a few "tricks," one of which is *chromatography.*

V. Just For Fun

Take a handful of salt and a handful of sugar and mix them in a bag. Now un-mix them. Draw or describe what you did.

Experiment 4

Making Goo

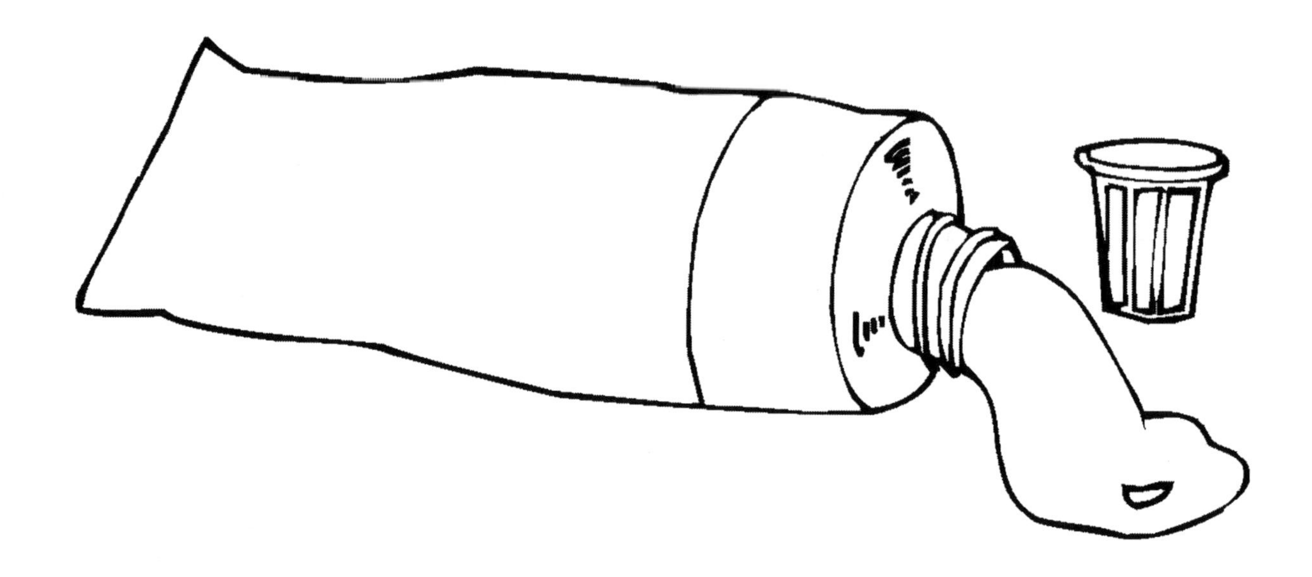

Introduction

Explore how the properties of substances can change when the substances are mixed together.

I. Think About It

❶ How does glue feel on your fingers? Can you roll it into a ball? Why or why not?

____

____

____

❷ How does laundry starch feel on your fingers?

____

____

____

❸ What do you think would happen if you added laundry starch to the glue?

____

____

____

❹ Do you think the glue would feel the same as it did before you added laundry starch to it? Why or why not?

____

____

____

5. What do you think would happen if you added glue to the laundry starch?

____

____

____

6. Do you think the laundry starch would feel the same as it did before you added glue to it? Why or why not?

____

____

____

II. Observe It

1. Take a small plastic cup and put about 30-60 ml (1/8-1/4 cup) of white Elmer's glue in it.

2. Now add 30-60 ml (1/8-1/4 cup) of liquid laundry starch to the Elmer's glue.

3. Mix the glue and starch with your fingers.

4. Pay careful attention to any changes that occur. Think about how the glue might be changing or how the starch might be changing.

5. Try to roll the glue and laundry starch into a ball. Are you able to do it?

____

____

III. What Did You Discover?

❶ What happened when you mixed the starch and glue together?

____

____

____

❷ Could you feel the glue and/or starch change?

____

____

____

❸ What did it feel like? Could you roll the mixture into a ball?

____

____

____

❹ Do you think this means there was a *chemical reaction?*

____

____

____

IV. Why?

You discovered that when you mixed glue and starch together, something happened. The glue became less sticky, and you found you could easily roll it into a ball.

When glue and laundry starch are added to each other, a *chemical reaction* occurs. You can see that a reaction has occurred because the glue and the laundry starch are different than they were before you mixed them together. The glue loses its stickiness and can be easily rolled into a ball.

The reaction you observed occurs between the molecules in the glue and the molecules in the laundry starch. Glue is made of long chains of molecules called *polymers*. The laundry starch hooks these polymers together by means of a chemical reaction. The reaction changes the way the polymers move around, and this changes the properties of the glue and the starch.

V. Just For Fun

Repeat the experiment using a different glue to see if it makes a difference in the results. Some ideas for glues are: blue glue, clear glue, wood glue, glitter glue, or paste glue. You can also try adding a little food coloring to your mixture.

In the box below, record your observations.

Experiment 5

Make It Rise!

Introduction

Discover more about molecules in food and the body.

I. Think About It

❶ List as many different types of molecules as you can.

❷ What kinds of molecules make food salty?

❸ What kinds of molecules are glue and starch made of?

❹ Do you think all the molecules in your body would work properly if your body got too hot? Why or why not?

❺ What kinds of molecules do you think make bread rise?

CHEMISTRY

II. Observe It

1. Open one package of active dry yeast. In a bowl, dissolve 1/2 of the yeast in 240 ml (1 cup) of lukewarm water. Label this **Dough A**. In a second bowl, dissolve the remaining half of the yeast in 240 ml (1 cup) of cold water and label it **Dough B**.

2. Add 15 ml (1 Tbsp.) of sugar and 15 ml (1 Tbsp.) of oil to each bowl and mix.

3. Add 475 ml (2 cups) of flour to each yeast mixture. Mix and knead.

4. Coat two bowls with oil. Place **Dough A** in one bowl and **Dough B** in the other bowl. Put the bowl with **Dough A** in a warm, moist place, and put the bowl with **Dough B** in the refrigerator.

5. Let the doughs rise for one to one and a half hours.

6. After the doughs have risen, carefully observe any differences.

7. Punch down each dough. Return **Dough B** to the refrigerator and let the doughs rise again for another hour.

8. Shape the doughs into loaves and bake at 190°C (375°F) for 35-40 minutes.

III. What Did You Discover?

❶ Describe what happened to the dough that was made with warm water and put in a warm place.

____

____

____

❷ Describe what happened to the dough that was made with cold water and placed in the refrigerator.

____

____

____

❸ Was there a difference between the two doughs?

____

____

____

❹ What was different about how you made the two doughs? Do you think this difference might have made the doughs behave differently?

____

____

____

IV. Why?

You should have discovered that the dough made with cold water and placed in the refrigerator did not rise. The dough made with warm water and placed in a warm place did rise. The dough made with cold water and placed in the refrigerator did not rise because yeast is made of living cells and living cells often need warm temperatures to live.

Yeast has molecules in it called *enzymes*. Enzymes produce the molecules that make bread rise. Many enzymes need warm temperatures in order to work properly. In fact, inside your body there are many enzymes that do a variety of jobs. Some enzymes cut molecules, some read other molecules, some copy other molecules, and some glue molecules together. All of the enzymes in your body work at your body temperature. If your body gets too hot or too cold, your enzymes can't work—just like the yeast enzymes could not make the molecules needed for dough to rise when it was made with cold water and placed in the refrigerator.

V. Just For Fun

Baking Powder (or Not) Biscuits

For this experiment see what happens when you make biscuits with and without baking powder. Before starting the experiment, record what you think will happen. Space is provided on the following page.

1. Take two mixing bowls and label one **Dough A** and the other **Dough B**.
2. Into each bowl measure 475 ml (2 cups) flour and 2.5 ml (1/2 teaspoon) salt. Mix together.
3. Add 15 ml (1 Tbsp.) of baking powder to **Dough A only**.
4. To each bowl add 60 ml (1/4 cup) soft butter. Mix with your fingers until the butter is well mixed in. The dough should look something like cornmeal.
5. To each bowl add 180 ml (3/4 cup) milk and mix until the dough particles cling together.
6. Knead each dough on a floured board for about 1 minute.
7. Take biscuit size pieces of dough and pat them into biscuits that are about 2 cm (3/4 inch) thick. Or roll out the dough to the same thickness and use a biscuit cutter.
8. Place each biscuit on an ungreased cookie sheet. Use a separate cookie sheet for each dough mixture.
9. Bake at 232°C (450°F) for 12–15 minutes until golden.
10. Record your observations of **Dough A** and **Dough B**.

CHEMISTRY

Baking Powder (or Not) Biscuits

Thoughts about what will happen

Dough A

Dough B

Results of the experiment

Dough A

Dough B

Experiment 6

Nature Walk

Introduction

The first and most important part of learning about plants is to observe them. What can you notice about plants?

I. Think About It

❶ What different types of plants have you observed?

❷ Do you have plants that live in your house? Are they different from each other? Why or why not?

❸ Do you have plants that live in your yard? Are they different from each other? Why or why not?

❹ What kinds of flowers have you seen on plants? How were they different from each other?

____

____

____

____

❺ Have you ever seen grass and weeds? How do you tell which is which?

____

____

____

____

❻ Have you noticed that some plants grow in one place but not in another? What do you think is different about the places where plants grow?

____

____

____

____

II. Observe It

1. Take a notebook, a pencil, some colored pencils, and a camera if you have one, and walk outside in a wooded area, a park, or in your yard—anywhere you can observe a variety of plants.

2. Locate two or more different kinds of plants you would like to study.

3. Draw the plants you see. Even if you are not a skilled artist, you can draw the basics of what the plants look like—for example, what size and shape the leaves are, whether there are few or lots of leaves, any flowers or seed pods or fruit they may have, how tall or short they are, and other features you observe. Also note the color of the plants.

4. Notice where the plants are growing, what the soil looks like, how much sun they are getting, the temperature, and how much water they have. Write details such as these in your notebook along with any other observations you'd like to record.

5. If you have a camera, take a photo, print it, and tape it in your notebook alongside your drawing.

III. What Did You Discover?

❶ Where were the plants you observed? In a field? In a pot? Near a river or stream?

____

____

❷ Did the plants have little, medium, or a lot of water?

____

____

❸ Did the plants have little, medium, or lots of sunshine?

____

____

❹ What was the surrounding temperature?

____

____

❺ How would you describe the kind of environment the plants you observed live in? (desert, wet or marshy, rocky, forest, etc.) What details did you observe?

____

____

____

____

IV. Why?

Observing plants and the environment they live in is the first step to learning about plants.

Making careful observations allows you to learn things such as how tall or short a plant is, what color the leaves or flowers are, whether or not it has thorns, and if it is fuzzy or smooth. By noticing the features of different plants, you will become aware of the wide variety of plants and how in some ways they are the same as each other and in some ways they are different. Biologists group plants into divisions based on careful observations of the features the plants have.

By making careful observations, you can also notice that different plants require different surroundings, or environments, in order to grow and be healthy. Plants require air, water, sunlight, and soil that has nutrients in it. But some plants can only live where the temperatures stay warm and there is not too much water. Other plants need freezing temperatures in the winter. Some plants need to live in the shade and others need to live where they get lots of sunlight and there is no shade. Some plants grow best when people take care of them, and most do just fine on their own.

The study of plants is called *botany*. Botanists have discovered many amazing things about plants and continue to make new discoveries.

V. Just For Fun

Make your notebook into a nature journal. Choose two or more different kinds of plants that you'd like to observe over a period of several months. You may want to observe the same plants as you did for the *Observe It* section, or you may want to choose different plants.

Ask yourself questions about the plants and their environment. How do the plants change—or do they stay the same? Do changes in things such as temperature and amount of rainfall effect the plants? Are bugs or animals eating them? Do the plants change with the seasons? Asking questions leads to more and better observations

Record your observations in your nature journal by drawing, writing about, and photographing the plants. You can also record anything else you find of interest in the plants' environment.

Experiment 7

Who Needs Light?

Introduction

What happens to plants when they don't get sunlight? Do this experiment to find out.

I. Think About It

❶ List three things a plant needs to have in order to live.

❷ What do you think would happen to a plant if it did not get light from the Sun?

II. Observe It

1. Take two small plants that are the same kind and about the same size.

2. Carefully observe each plant. Note any unique features they have.

3. Make a list of words that describe the plants and their features.

❹ Label one plant **A** and the other **B**. Draw each plant.

A	B

❺ Take the plant labeled **A** and put it in a sunny place.

❻ Take the plant labeled **B** and put it in a dark place.

❼ Describe what you think will happen to each plant.

Plant A

____

____

____

____

____

____

Plant B

____

____

____

____

____

____

❽ Make a schedule for watering your plants on a regular basis. Be sure to water both plants with the same amount of water.

Draw each plant after week 1.

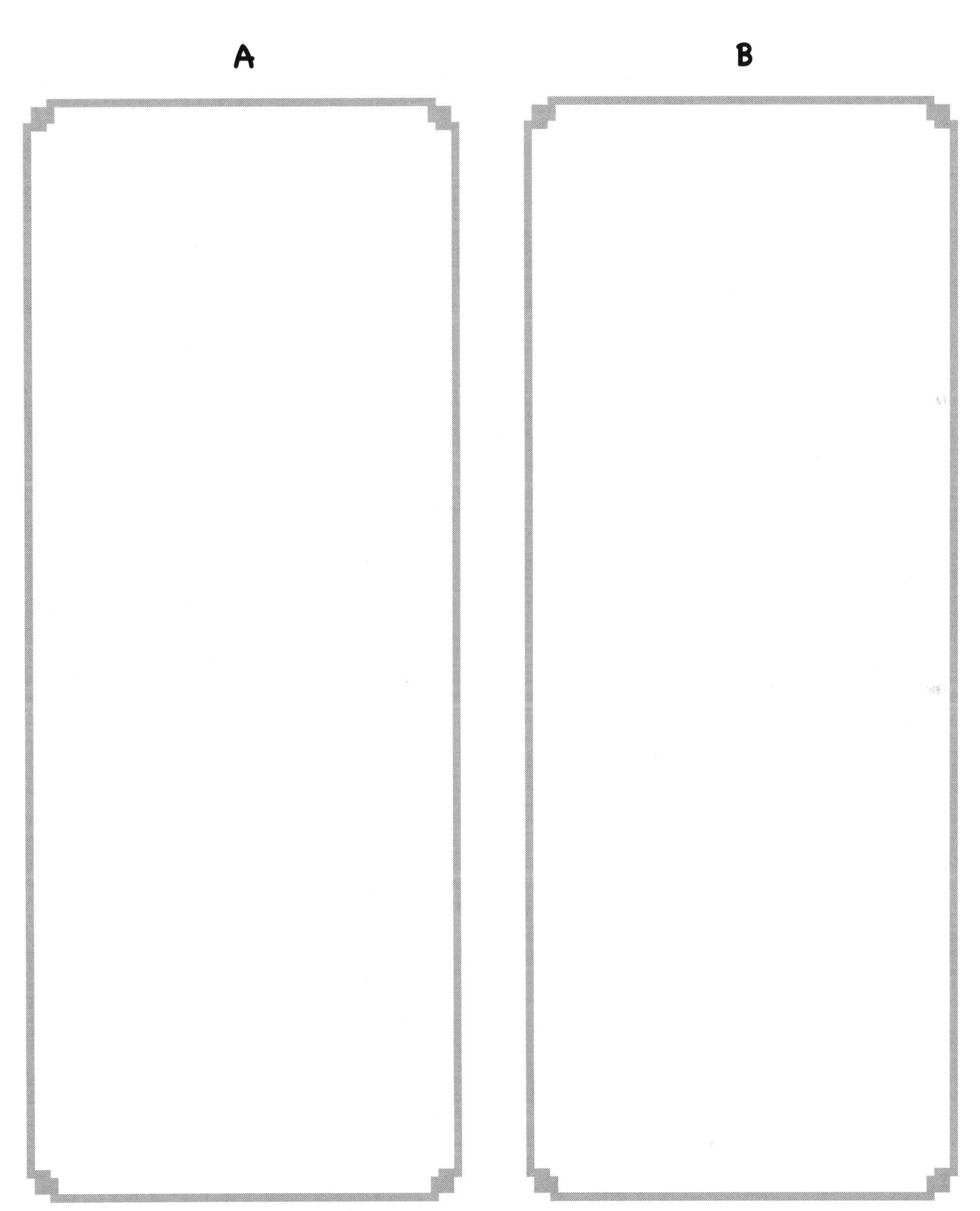

Draw each plant after week ________

Draw each plant after week ________

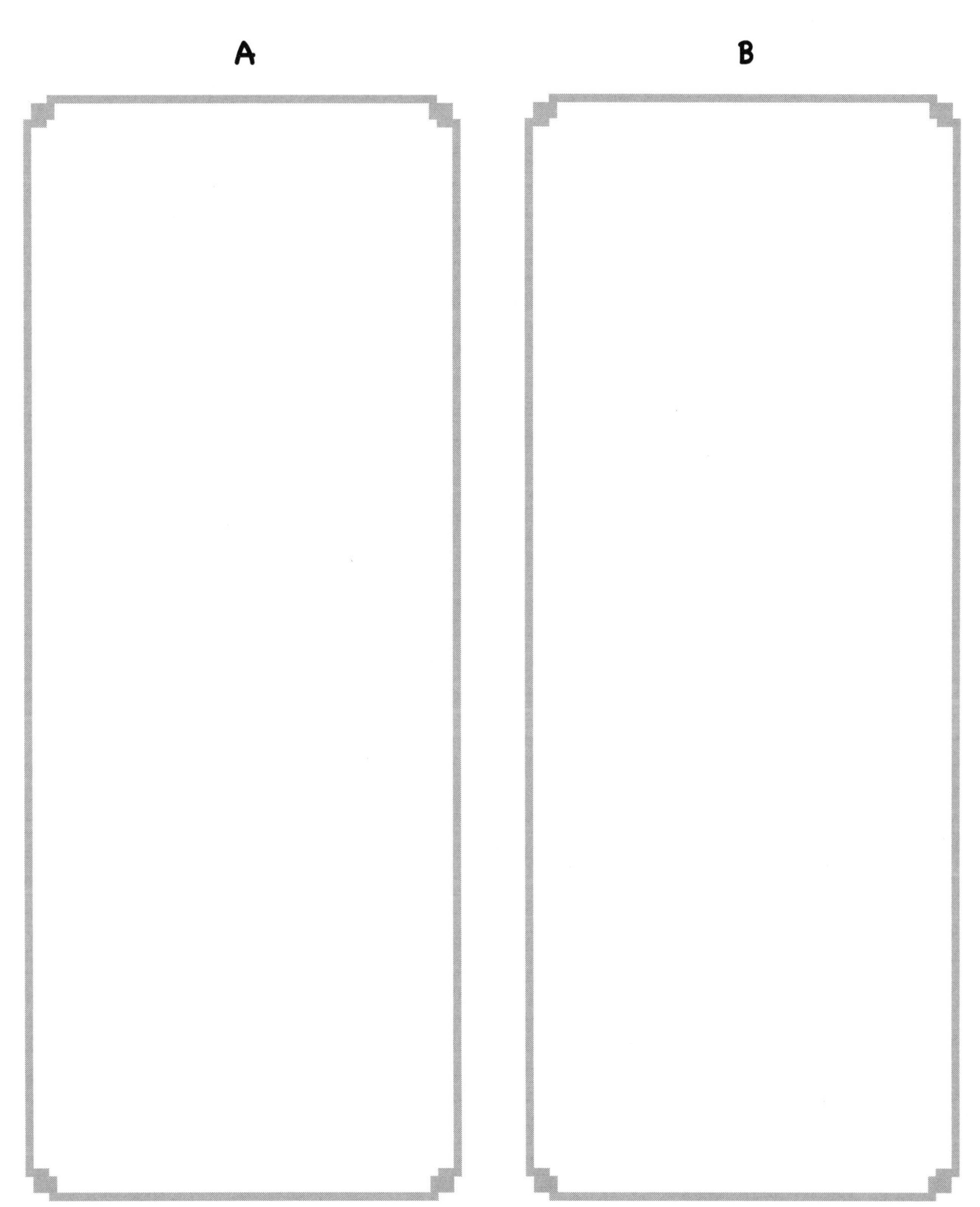

Draw each plant after week ________

A

B

III. What Did You Discover?

❶ How did the plants look on the first day?

Plant A ______________________________

____

Plant B ______________________________

____

❷ How did the plants look after the first week?

Plant A ______________________________

____

Plant B ______________________________

____

❸ How did the plants look after the last week?

Plant A ______________________________

____

Plant B ______________________________

____

❹ Describe any differences you observed between the two plants.

Plant A ______________________________

____

Plant B ______________________________

____

IV. Why?

A regular houseplant needs sunlight to make food. If a houseplant is not able to get sunlight, it cannot make the food it needs to stay healthy. Eventually, a houseplant will die if it does not get enough sunlight.

When you put one plant in the dark and keep one plant in the sunlight, you are testing what happens to a plant that does not get sunlight. Why do you think you needed two plants—one in the sunlight and one in the dark?

You used two plants because, as a scientist, you want to make careful observations when you change something. When you use two plants (one in the sunlight and one in the dark), you can easily compare any changes in each of the plants. You want to know what happens to a plant that is in the dark compared to a similar plant that stays in the sunlight.

This is called using a *control*. The control shows you what will happen if nothing is changed. In this way, scientists can be sure that they will be able to observe what happens when something is changed. In this experiment, you observed what happened when sunlight was taken away from a plant. Your control plant (plant **A**) showed you what the plant looked like when it had sunlight. The plant you took away from the sunlight (plant **B**) showed you what happened to the plant when it could not use the Sun's energy to make food. Using a control helped you determine what happens when a plant does not get sunlight.

V. Just For Fun

Do another experiment using two more plants that are both the same kind and about the same size. This time water one plant and don't water the other one. Based on the last experiment, think of the steps you will need to take to preform this experiment. Record your results at the beginning and the end of the experiment.

Plant A — Beginning **Plant B — Beginning**

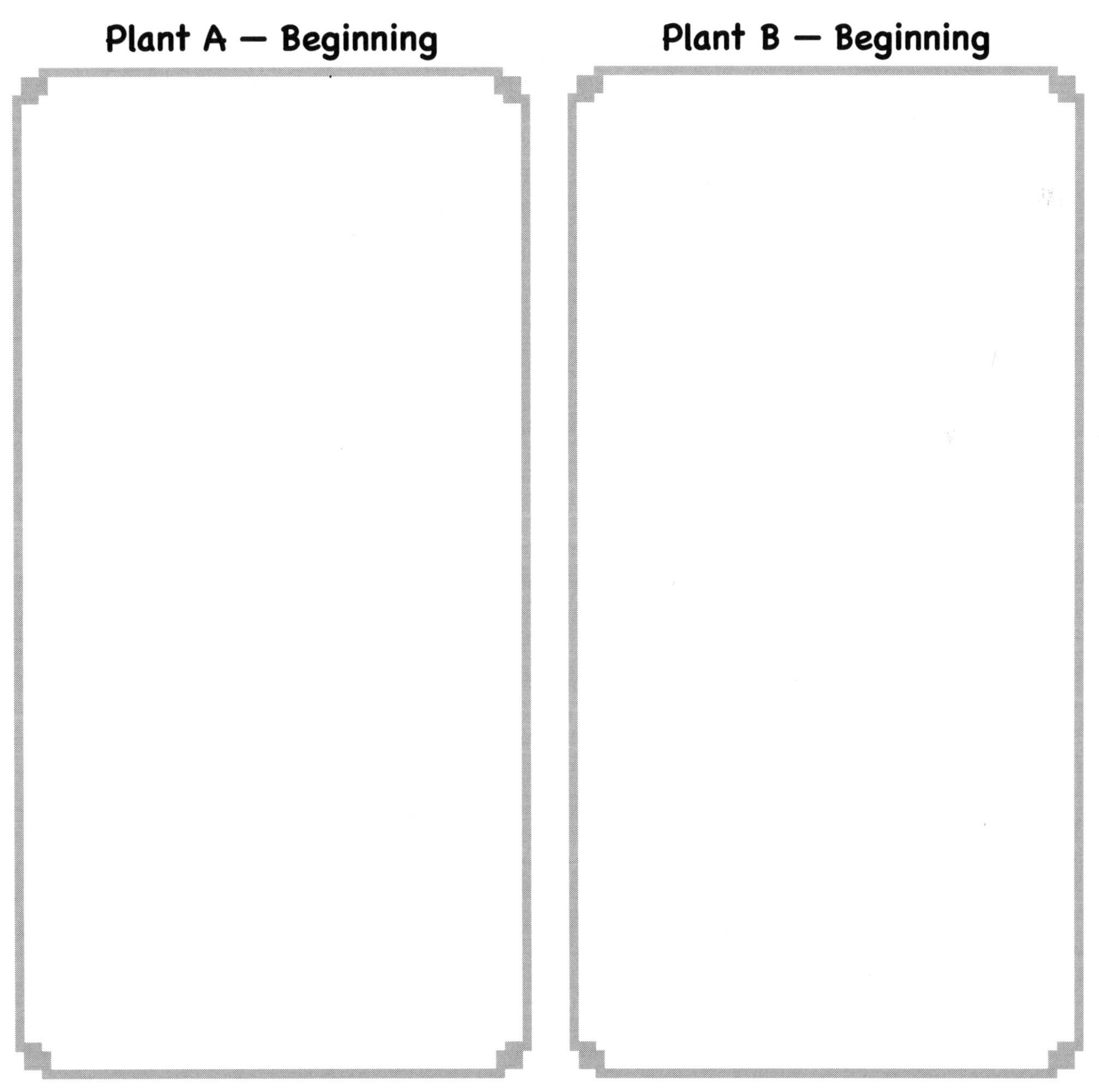

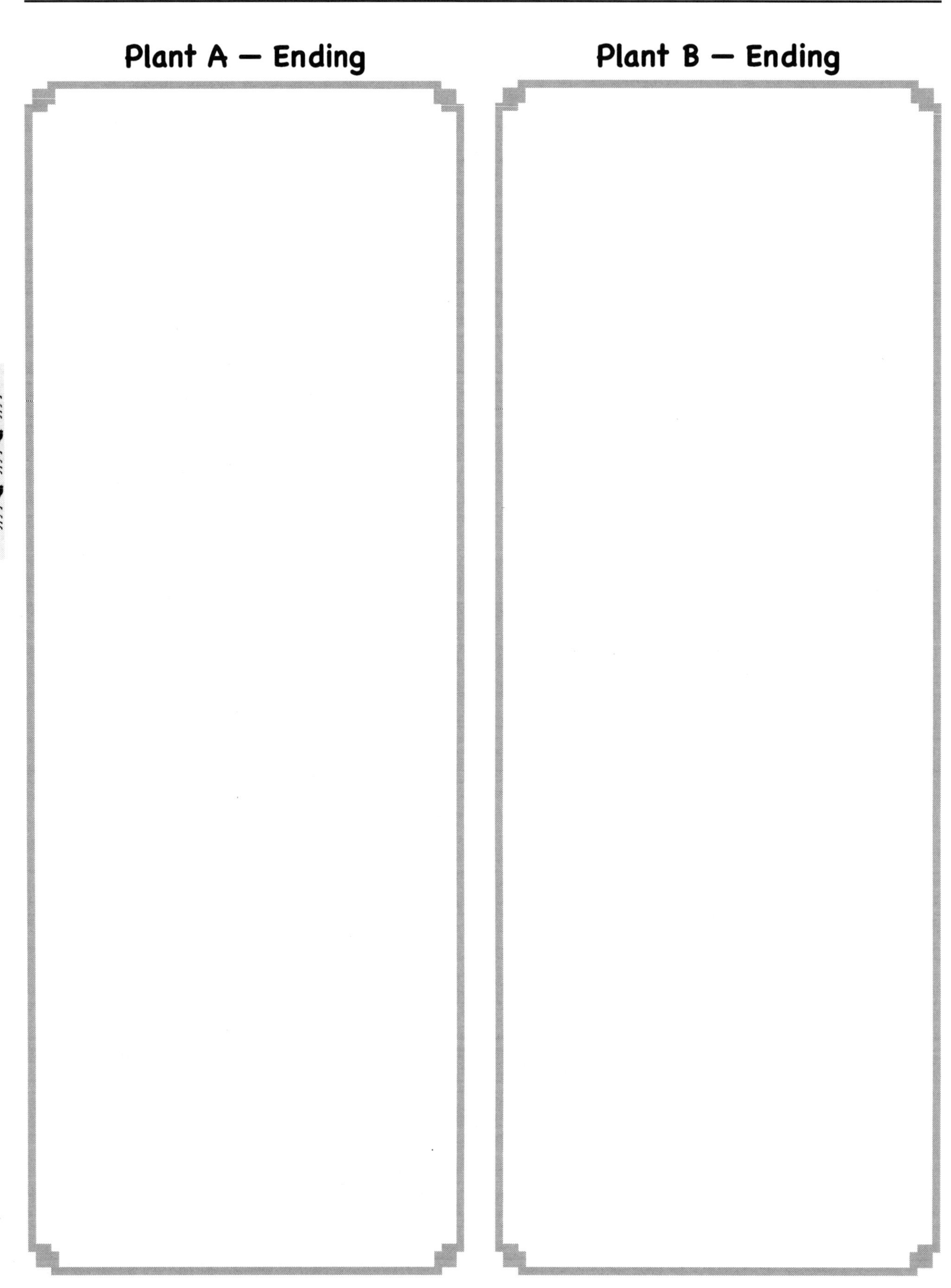
Plant A — Ending
Plant B — Ending

Experiment 8

Thirsty Flowers

Introduction

How does water travel through a flower? This experiment will help you find out.

I. Think About It

❶ If you put a white carnation into a glass of colored water, what do you think will happen to the flower?

____

____

____

____

❷ Draw a picture showing what you think will happen.

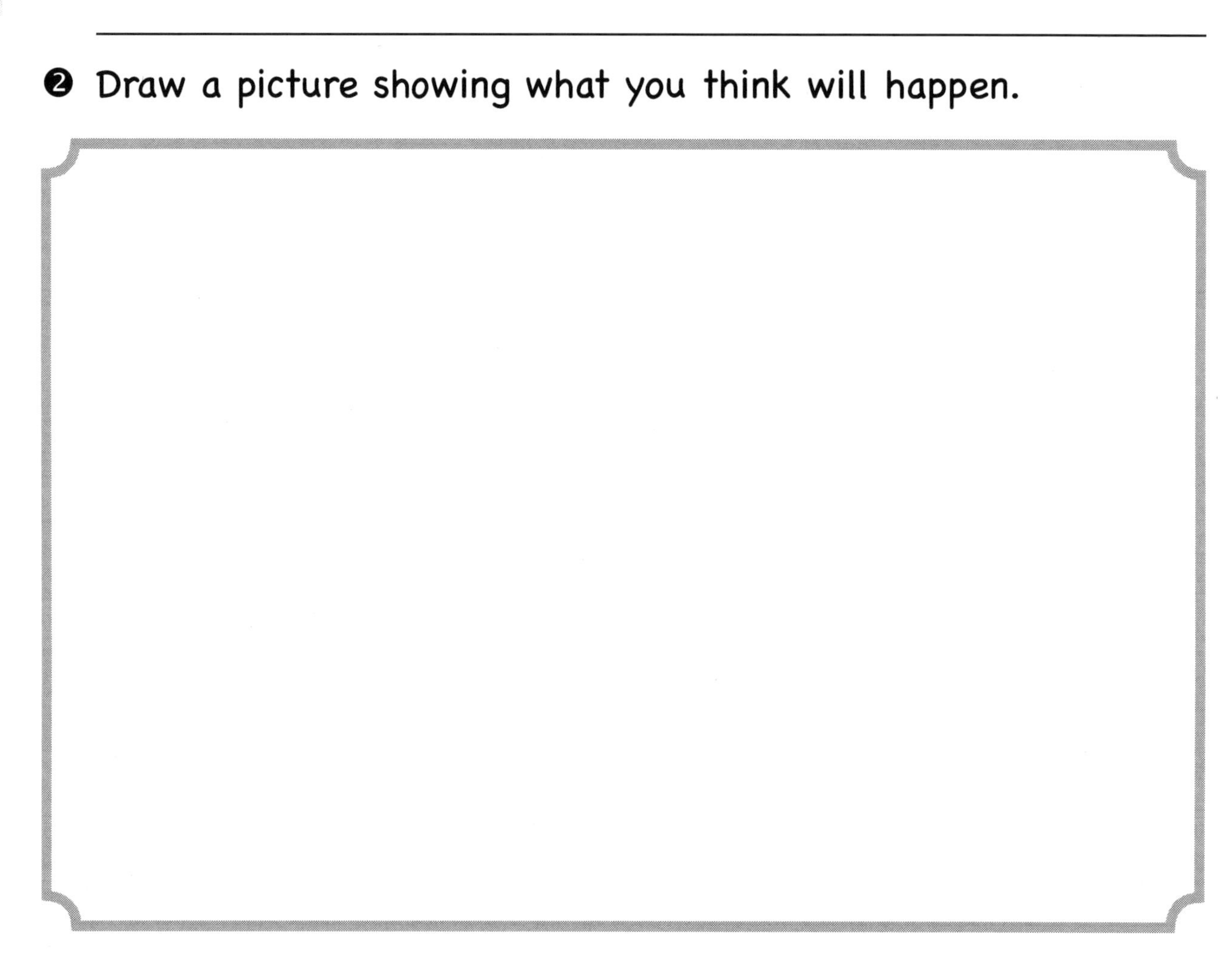

II. Observe It

❶ Carefully observe the carnation. Draw your observations.

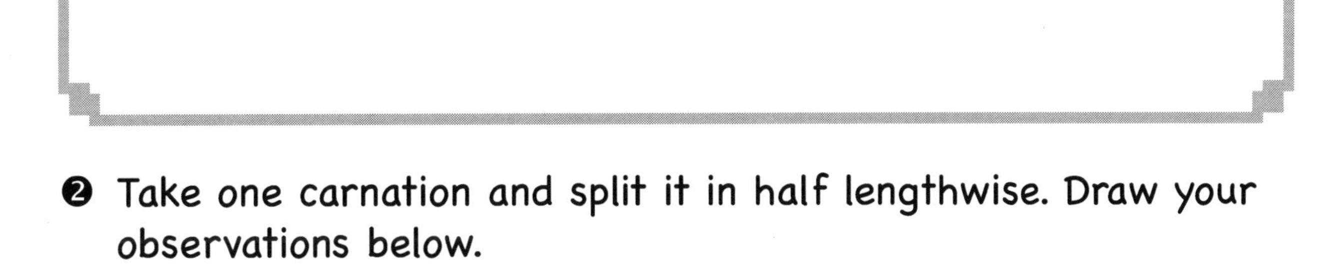

❷ Take one carnation and split it in half lengthwise. Draw your observations below.

❸ Take another carnation and place the stem in colored water. Observe what happens to the flower. Use the next boxes to draw what you see happening over the next several minutes.

After ________ minutes.

After ________ minutes.

After ________ minutes.

❹ Cut the stem open. Draw what you see.

III. What Did You Discover?

❶ What did the carnation look like before you added the colored water?

____

____

____

❷ What did the carnation look like inside?

____

____

____

❸ What happened to the carnation when you put it into the colored water?

____

____

____

❹ What did you observe in the stem after it was in the colored water?

____

____

____

IV. Why?

A carnation is a flower that usually has a long green stem. The stem is the part of the plant that carries water and nutrients up the plant. The roots of the plant take water and nutrients from the soil, and tissues in the stem then act like little straws that draw water from the roots up the plant.

When you place a carnation in colored water, you can observe what happens to the flower. Because the water is colored, you can watch the flowers "drink" the water from the glass. The colored water travels up to the top of the stem of the carnation, and when it reaches the flower, it starts to color the petals.

When you cut open the carnation stem, you can observe the tissues that move the water up the stem through the plant. With the stem cut open, you may be able to observe that the tissues in the stem are colored too.

These tissues in a carnation stem are designed to move the water one way—up the stem. The water does not come back down through the stem. (Do you think that is true? Why don't you try it—take the stem out of the colored water, and see if the colored liquid runs back out.)

A plant "drinks" water much like you drink water from a straw. Can you explain how a plant "drinks" water?

V. Just For Fun

Repeat the experiment with a different kind of white flower. Observe and record how long it takes for the flower to become colored. Did anything different happen with this flower than with the carnation? Record your observations below, noting similarities and differences.

Experiment 9

Growing Seeds

Introduction

This experiment will help you find out what happens as a seed grows into a plant.

I. Think About It

❶ If you put a dried bean in a jar, add some water, and then let it sit for several days, what do you think will happen?

____

____

____

____

❷ Draw a picture showing what you think will happen.

II. Observe It

❶ Carefully observe a dried bean. Look at the outside. Draw your observations.

❷ Take the bean and split it in half lengthwise. Draw your observations.

3. Take a clear glass jar and a piece of absorbent white paper. Cut a piece of the paper so that it is long enough to go all the way around the jar. Then wrap the piece of paper around the inside of the jar.

4. Place two dried beans between the paper and the jar. The paper should hold the beans against the side of the jar, but if it doesn't, you can tape the beans to the jar.

   Make sure the beans are not touching the bottom of the jar but are placed about 6-12 mm (1/4-1/2 inch) above the bottom.

5. Pour some water in the bottom of the jar so that the water contacts the absorbent paper but not the beans.

6. Place plastic wrap on top of the jar and fasten it with a rubber band to seal the jar and prevent evaporation of the water.

7. In the box on the next page, draw the beans and the jar. Include any details you observe.

Observations: Day 1

8. Check your beans frequently to observe their growth. As you notice changes, record your observations in the following boxes. Add water to the jar as needed to make sure the paper stays moist.

Observations: Day ______

Observations: Day ________

Observations: Day ______

Observations: Day ______

Observations: Day _______

III. What Did You Discover?

❶ What did the inside of the dried bean look like when you opened it? What did you find inside?

❷ How many days did it take for the beans to start growing?

❸ Which part started to grow first? Which way did it grow—up or down?

❹ How many days did it take for the beans to turn into seedlings? In your own words, describe briefly how they grew from bean to seedling.

IV. Why?

A bean is a seed. Seeds are how most plants begin. Inside a bean you can see the embryo that will grow into the little plant, or seedling. Inside the bean you can also see the food the embryo uses to grow until it has the roots and leaves it needs to make its own food.

When you put a bean in the ground, it will start to sprout. You can watch a bean sprout by putting it into a clear jar and adding water. The bean starts to sprout a root first. The root will grow downward, finding its way to the ground. A root knows which direction to grow, and it will not grow upward toward the Sun but down into the ground. The shoot of the plant will grow next. It will grow upward toward the Sun so that when the leaves come out, they can collect the sunlight.

The bean continues to grow the roots and the shoot until it becomes a seeding. When the seedling has leaves and a root big enough to gather nutrients, it no longer needs the food it had inside the seed. It is ready to become a big plant!

V. Just For Fun

Repeat the experiment using a different kind of seed. You might try a different type of dried bean or pea seed. You might try garden seeds you get at the store. Or you might gather seeds from a raw fruit or vegetable you are eating, such as a tomato, cucumber, watermelon, or squash. If you gather the seeds yourself, let them dry before putting them in the jar. Record your observations.

Observations of ____________________ Seeds

Experiment 10

Lemon Energy

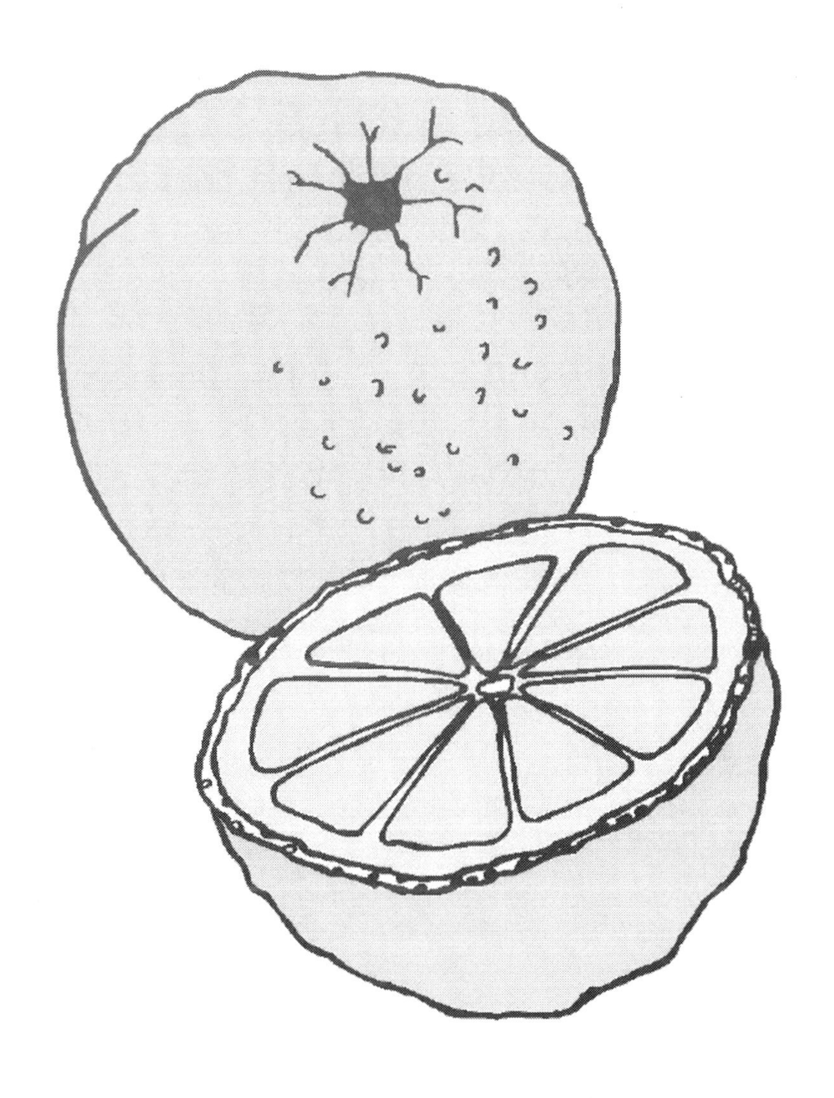

Introduction

Do you think lemons have stored chemical energy? Explore that idea in this experiment.

I. Observe It

❶ With the help of an adult, take three lemons and stick a copper penny in one end of each lemon and a zinc nail in the other end.

❷ Connect the three lemons together using copper wire and alligator clips or duct tape. Connect the penny side of one lemon to the zinc side of another lemon.

❸ Take the small LED light and clip or tape the free end of one of the wires to one end of the LED. Clip or tape the free end of the other wire to the other end of the LED.

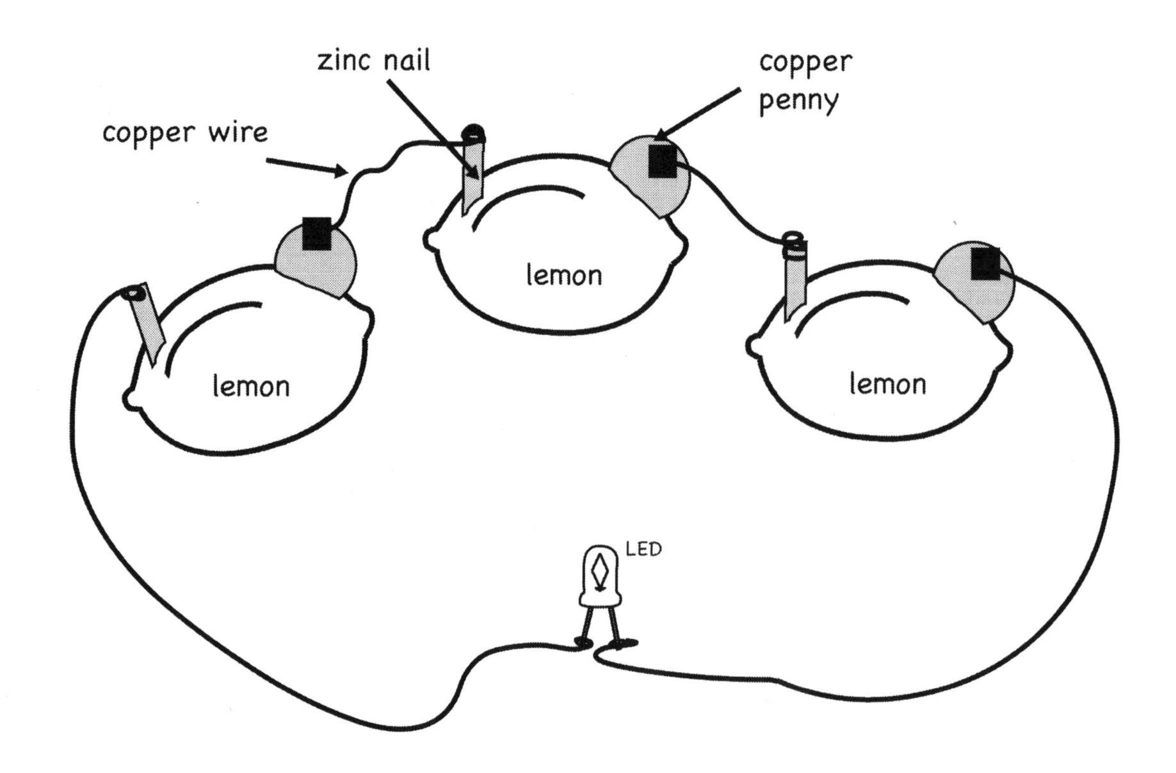

4. Observe what happens when you attach the LED to the free ends of the wires.

5. Record your observations below.

❻ Observe what happens to the LED when you disconnect one of the wires attached to a penny.

❼ Record your observations below.

8. Reconnect the wire. Now observe what happens to the LED when you disconnect one of the wires attached to a zinc nail.

9. Record your observations below.

Summarize Your Observations

Trial	What Happened?
All Wires Connected	
Penny Wire Disconnected	
Zinc Wire Disconnected	

II. Think About It

❶ Think about how you set up the lemon battery. Think about how you connected the wires, how you connected the LED to the lemons, and what happened then.

❷ Review what you learned about chemical energy and batteries in Chapter 10 of the textbook.

❸ Based on your observations, which of the following statements is true? Draw a circle around it.

The LED will light only when all the wires are connected.

The LED will light with the penny wire disconnected.

The LED will light when the zinc wire is disconnected.

❹ Think about any problems you might have had with your experiment. List them below.

___

___

___

___

___

___

___

III. What Did You Discover?

1. Were you able to get the LED to light up? Why or why not?

2. Did the LED light up when one or more of the wires was disconnected? Why or why not?

IV. Why?

Before you did this experiment, did you know a lemon could be a battery? We normally don't think of lemons as batteries but as food. However, in this experiment you assembled an electric circuit using lemons as batteries. The lemons contain acid. The acid in the lemons reacts with the copper and zinc metals, and this creates a chemical reaction. This chemical reaction inside the lemons produces electricity that can be used to power a small LED. [The term LED stands for "light emitting diode" which is like a little light bulb that only requires a small amount of electricity to light up.]

Three lemons are used to light the LED. One lemon would not generate enough electricity to run the LED, so three lemons are needed. The lemons were connected to each other in such a way that the electricity of all the lemons could be added together. When all three lemons are working together, there is enough electricity generated by the chemical reactions of the combined lemons to light up the LED.

When one or more of the wires is disconnected, the LED stops working. By disconnecting the wires, the battery energy in each lemon can no longer be added to that of the other lemons, the electricity cannot flow to the LED, and so the LED stops working. The lemons are a form of stored chemical energy, and once they are connected to each other in the right way, they can generate enough electricity to power a small LED.

V. Just For Fun

Disconnect one of the wires from the LED. With the fingers of one of your hands, hold the wire connected to the lemons. With the fingers of your other hand, hold the metal end on the disconnected side of the LED.

What happens?

Experiment 11

Sticky Balloons

Introduction

With this experiment, see if you can find out how much charge a balloon has.

I. Observe It

❶ Take a rubber balloon and blow it up with air. Tie the end closed, then place the balloon on a wall. Observe whether it sticks to the wall.

❷ Without popping the balloon, rub it in your hair.

❸ Carefully pull the balloon away from your hair and observe whether your hair sticks to the balloon. If your hair sticks to the balloon, continue to Step ❹. If your hair does not stick to the balloon, rub the balloon in your hair again.

❹ Test how sticky the balloon is by placing it on a wall. Observe whether the balloon sticks or falls off the wall.

❺ Record your observations in the space provided on the next page. Use the following questions.

- Does the balloon stick?
- How long does the balloon stick? 1 second? 2 seconds? 10 seconds? Longer than a minute?
- Does the balloon move around or stay still?
- What happens if you blow gently on the balloon? Does it stay stuck or does it fall off?

Hair

❻ Repeat Steps ❷-❺. This time, instead of rubbing the balloon in your hair, rub it on different materials, such as wool or cotton clothing, metal or wood surfaces. In the following boxes record your observations for each object.

PHYSICS

PHYSICS

PHYSICS

II. Think About It

1. Think about the balloon and the different materials you used to charge the balloon.

2. Review what you learned in the textbook about electrons, charges, and force.

3. Create a chart below that lists the materials or surfaces you used in your experiment. Begin with those materials or surfaces that created the most charge, and end with those that created the least charge.

Most Charge ↓ Least Charge	

III. What Did You Discover?

❶ Were you able to get the balloon to carry a charge? Why or why not?

____

____

____

____

❷ Did some materials charge the balloon more than other materials, or were they all the same?

____

____

____

____

❸ Could you tell how much charge the balloon gained by attaching it to the wall? Why or why not?

____

____

____

____

IV. Why?

In this experiment you explored how a balloon can pick up electrons from other objects. When a balloon picks up electrons from another object, the balloon becomes charged. The very first time you placed the balloon on the wall, before you rubbed it in your hair, the balloon likely fell off. Why? It fell off because the balloon did not carry any additional charges.

When you rubbed the balloon in your hair, the balloon picked up electrons from your hair. The electrons are negatively charged, so the balloon became negatively charged. A negatively charged balloon will stick to surfaces that are slightly positively charged. If the balloon stuck to the wall, then the wall was slightly positively charged.

You could test how many electrons the balloon picked up by observing how easily the balloon would stick to a wall after it was charged. If the balloon stuck a lot, then there were lots of electrons picked up. If the balloon only stuck a little, then there were fewer electrons.

It is possible that it was difficult to get the balloon to become charged no matter what material or surface was used. If you discovered this, that's OK. Check the humidity in your area. If it was humid when you did the experiment, the electrons could not stay very long on the balloon.

V. Just For Fun

Take two balloons and tie a piece of string to the end of each balloon. Tie the other ends of the strings together and hang the balloons from a doorway or shower rod. What happens?

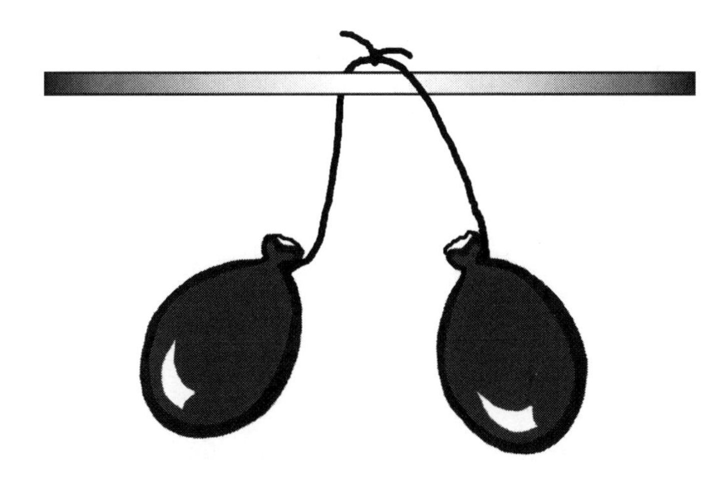

Take the balloons and rub them both in your hair. Let the balloons go and allow them to float back together. What happens?

Record your observations below.

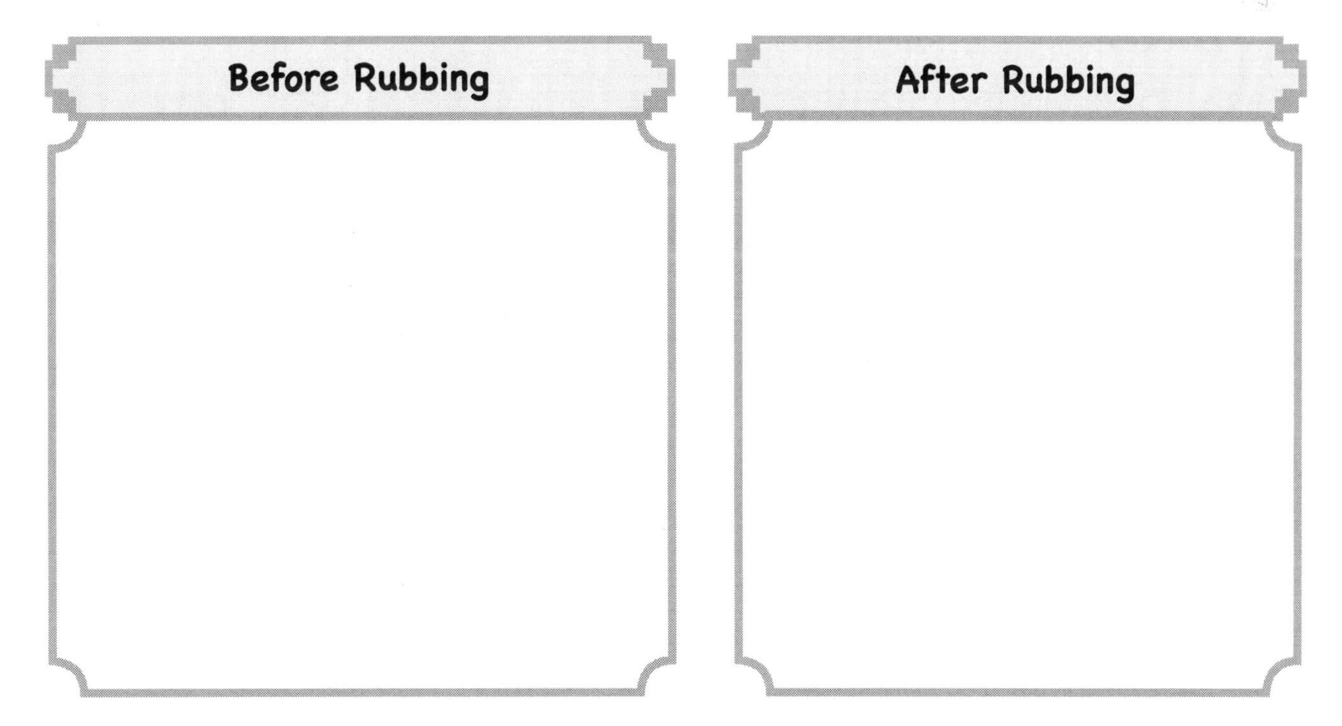

Experiment 12

Moving Electrons

Introduction

Discover whether using different materials can affect the flow of electrons in a moving electric current.

I. Observe It

1. With the help of an adult, set up the lemon battery from Experiment 10. Make sure the LED is illuminated.

2. With the help of an adult, take one of the wires between two lemons and cut it. Observe what happens to the LED. Write or draw your observations below.

Wires Apart

❸ Reconnect the ends of the wire. Observe what happens to the LED. Write or draw your observations below.

Wires Connected

❹ Disconnect the ends of the cut wire and place a piece of Styrofoam between them. Observe what happens to the LED. Write or draw your observations below.

Styrofoam

❺ Remove the Styrofoam from between the wires. Reconnect the wires by twisting them together. Observe what happens to the LED. Write or draw your observations below.

Wires Connected

❻ Repeat Steps ❹-❺ using the different materials on the following pages. Test all of the materials listed. Record your observations in the space provided.

Plastic Block

Cotton Ball

PHYSICS

Nickel Coin

Metal Paper Clip

Plastic Paper Clip

❼ Summarize your results below. Write "ON" or "OFF" in the LED column for each of the items listed.

Item	LED
Start—wires connected	
Wires apart	
Wires connected	
Styrofoam	
Wires connected	
Plastic block	
Cotton ball	
Nickel coin	
Metal paper clip	
Plastic paper clip	

II. Think About It

❶ Think about the LED and the different materials you placed between the wires.

❷ Review what you learned in your *Student Textbook* about moving electric charges.

❸ Use the chart below to organize the test results for the various materials. Place those items that illuminated the LED in one column and those items that did not illuminate the LED in the other column.

LED "ON"	LED "OFF"

III. What Did You Discover?

❶ Which items illuminated the LED?

❷ Which items did not illuminate the LED?

❸ Were the items that illuminated the LED metals?

❹ Were the items that did not illuminate the LED non-metals?

❺ Why do you think metals illuminated the LED and non-metals did not?

IV. Why?

In this experiment you explored how electrons moved (or didn't move) through different materials. When the lemons in the battery are connected with metal wires, the electrons can flow freely through the wires and light up the LED. When the metal wire is cut, the electrons are stopped from flowing through to the LED. When the metal wires are reconnected, the electrons flow through the wire again to light up the LED.

You discovered that some materials will allow electrons to flow through them, and some materials won't. Electrons easily flow through most metals. Metals and any other materials that allow electrons to flow through them are called *conductors*. Electrons do not flow through most plastics (Styrofoam, plastic blocks, or plastic paper clips), cotton balls, and other similar materials. These materials are called *insulators*.

There are also materials that are reluctant to allow electrons to flow through them. These materials are called *resistors*. Resistors will allow a few electrons to flow through them, but not all electrons. Resistors are used in electronic circuits to control the amount of electron flow.

V. Just For Fun

Take the two ends of the wire that was cut. Place both ends in a glass of water without letting them touch each other. Observe what happens to the LED.

Now add 15 ml (1 tablespoon) of salt to the water. Stir the salt until it has completely dissolved. Again place the wire ends in the glass and observe what happens to the LED.

Record your observations below.

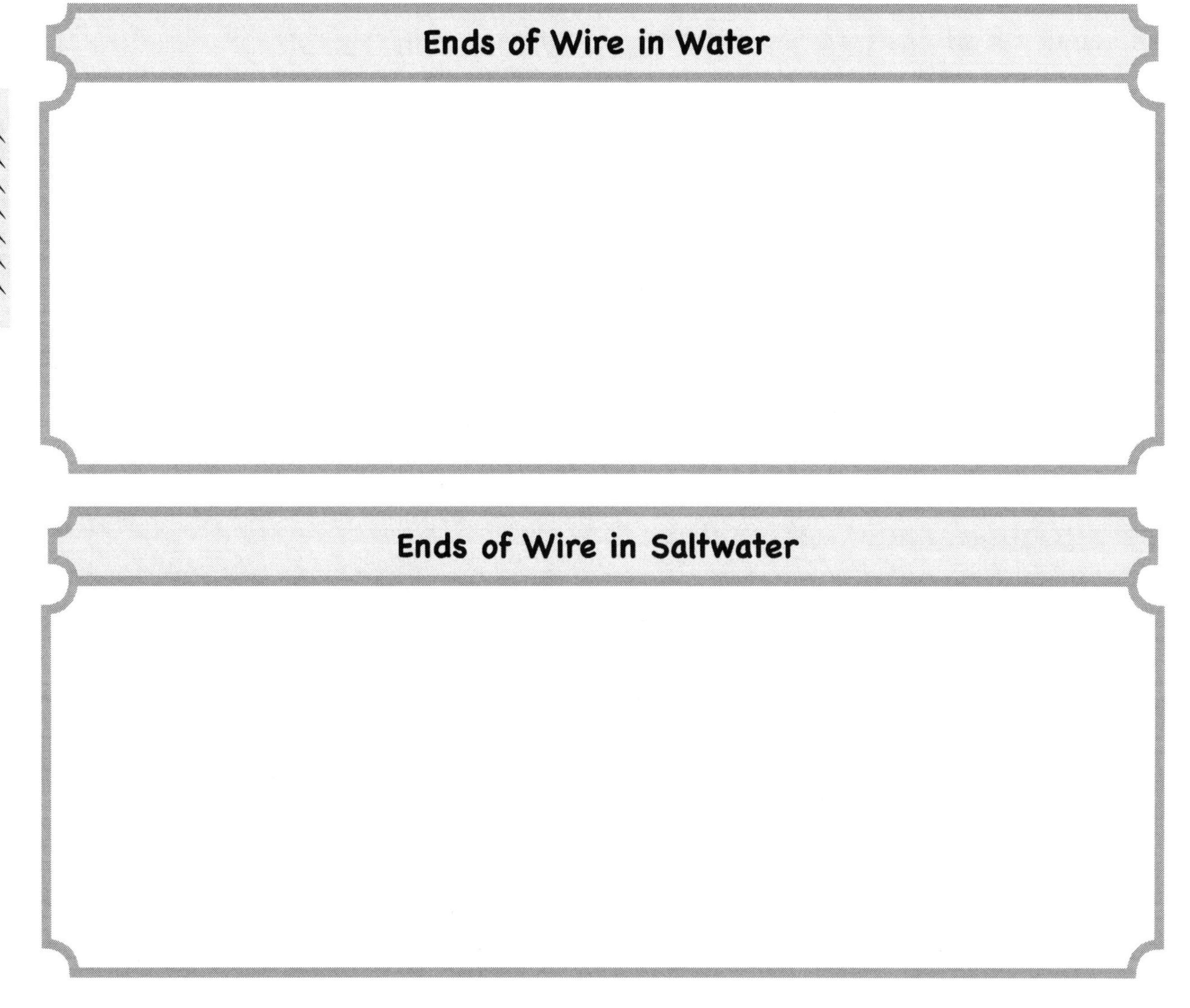

Experiment 13

Magnet Poles

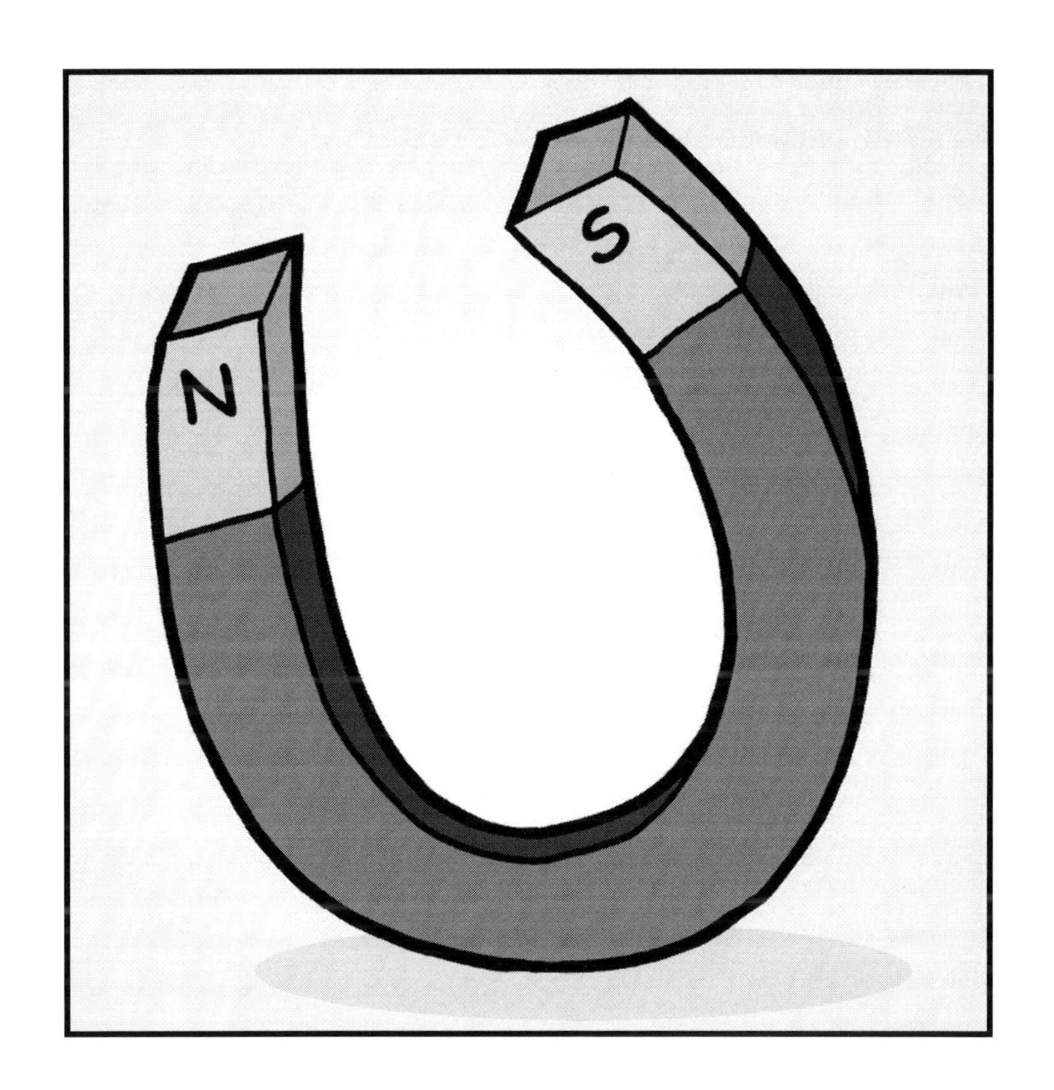

Introduction

Explore how magnet poles behave!

I. Observe It

❶ Take two magnets and place them on a table several inches apart with each "**N**" facing the other.

❷ Gently push one "**N**" closer to the other "**N.**" Observe what happens. Write or draw your observations below.

Trial 1

❸ Place the two magnets on the table several inches apart. Reverse the direction of one of the magnets so that the "**S**" of one is facing the "**N**" of the other.

❹ Gently push the "**N**" closer to the "**S**." Observe what happens. Write or draw your observations below.

Trial 2

❺ Repeat Steps ❶-❹ several times. See how close you can bring the two magnets together before something changes.

Write or draw your observations in the spaces provided.

Trial 3

Trial 4

Trial 5

PHYSICS

Trial 6

Trial 7

Trial 8

Trial 9

Trial 10

PHYSICS

❻ Summarize your results in the chart.

➟ Mark the trials **N-N** or **N-S**.

➟ Mark the trials where the magnets came together.

➟ Mark the trials where the magnets pushed apart.

Trial	N-N or N-S	Together or Apart
Trial 1		
Trial 2		
Trial 3		
Trial 4		
Trial 5		
Trial 6		
Trial 7		
Trial 8		
Trial 9		
Trial 10		

II. Think About It

❶ Think about the magnets and the different ways you pushed the magnets together.

❷ Review what you learned in your *Student Textbook* about magnets and magnetic poles.

❸ Look at the results you gathered in the previous section. Cover up the second column of the table and look only at the first and third columns. Without looking at the second column, write down those trials where the poles were the "same" and those trials where the poles were "opposite."

Trial	Same or Opposite
Trial 1	
Trial 2	
Trial 3	
Trial 4	
Trial 5	
Trial 6	
Trial 7	
Trial 8	
Trial 9	
Trial 10	

❹ Uncover the second column from the previous page. Do your answers match the second column of the previous table?

III. What Did You Discover?

1. What happened when you pushed the two "**N**" ends of the magnets together?

2. What happened when you reversed one of the magnets and pushed them together again?

3. If you had two magnets where the poles were not labeled "**N**" and "**S**," could you guess which poles were the same and which were opposite? Why or why not?

IV. Why?

In this experiment you explored magnetic poles. Magnets have two poles, one called "north" and one called "south." When the two opposite poles (north and south) come together, they attract each other, and the magnets will snap together. When two of the same pole (north and north, or south and south) come together, they repel each other, and the magnets will move away from each other.

Even though you may not know which pole is "north" and which pole is "south," by "playing" with the magnets (by reversing one of the magnets on the table and then switching it back), you can explore how the different poles react to each other. Reversing the magnet several times gives you information about when the magnets are coming together and when they are moving apart. As you observe the magnets coming together or moving apart, you are observing the effects of the different poles and can tell which poles are the same and which are opposite.

Scientists have to "play" with things around them to figure out what is happening. Scientists do different trials, just like you did, to find out what happens when some part of an experiment is changed. By "playing" with their experiments, scientists make observations they might have missed if they did the experiment only one way.

V. Just For Fun

Take a magnet and find out which surfaces in your house the magnet will attract or not attract.

Record your observations below.

Surface	Attract or Not Attract

Experiment 14

How Fast Is Water?

Introduction

Do you think anything different will happen if you pour water on sand or pour it on pebbles? Find out here!

I. Think About It

1. When rain falls on the land, where do you think the rainwater goes?

2. Does the kind of ground that rain falls on affect what happens to the rainwater? Why or why not?

3. Do you think there is water under the ground? Why or why not?

❹ What do you think would happen to plants and animals if it did not rain? Why?

____

____

____

____

❺ Why do think rain makes mud puddles in some places but not others?

____

____

____

____

II. Observe It

❶ Take three Styrofoam cups and use a pencil to poke a hole in the bottom of each one.

❷ Measure about 240 milliliters (1 cup) of sand and pour it into one of the cups.

❸ Measure about 240 milliliters (1 cup) of pebbles and pour them into the second cup.

❹ Measure about 240 milliliters (1 cup) of small rocks and pour them into the third cup.

❺ Next, measure about 120 milliliters (4 ounces) of water and pour it into the cup containing the sand. Observe how long it takes the water to run through the sand.

❻ Write your observations in the box provided. In addition to length of time, record anything else you observe.

❼ Repeat Steps ❺ & ❻ with the pebbles and then with the small rocks. Compare how long it takes the water to run through each cup.

How Quickly or Slowly Water Runs

SAND	
PEBBLES	
SMALL ROCKS	

III. What Did You Discover?

❶ Which material did the water run through the fastest? Why?

____

____

____

❷ Which material did the water run through the slowest? Why?

____

____

____

❸ Do you think all of the water came out of the cup with the sand? Why or why not?

____

____

____

❹ Do you think all of the water came out of the cup with the small rocks? Why or why not?

____

____

____

IV. Why?

In this experiment the water ran through the pebbles and the small rocks faster than it did through the sand. This happens because the particles of sand are much smaller than the pebbles and rocks. The small size of the sand particles allows them to fit together closely, leaving less empty space for the water to flow through. The pebbles and rocks have bigger spaces between the pieces, so water can flow through more quickly. The amount of empty space between the particles of a material is called its *porosity*.

The type of ground that rain falls on affects how much water can go into the ground. If rain falls on ground made of pebbles or small rocks, the rainwater will quickly flow through it. Rainwater also flows fairly quickly through sand and may not leave much water on the surface for animals to drink.

Another kind of soil is called clay. Clay has very small particles with little space between them. It can take water a long time to go through clay. Rainwater tends to form puddles on this type of soil. Clay can make very slippery and sticky mud!

The best type of soil for plants is in between sandy soil and clay soil and contains matter from decayed plants and animals. This kind of soil allows rainwater to flow through it slowly, keeping enough water for plants to use for making food.

V. Just For Fun

Mud City!

Make a mud city using dirt, water, pebbles, and rocks. Using what you learned in the previous sections of this experiment, make rivers and a lake that holds water.

You can use pebbles, rocks, sand, or dirt for city walls or to contain the rivers and lake. Think about how you can make the water stay in the lake and what will make water flow in rivers. Can you add things that will make the water flow faster or slower or change direction?

Once you've built your city, you can make paper boats to float on the rivers and the lake.

On the next page you can draw the mud city you created.

My Mud City

Experiment 15

What Do You See?

Introduction

If you explore the area where you live, do think you can observe things you haven't noticed before?

I. Think About It

❶ What living things do you think you will see if you walk around your yard and your neighborhood?

____

____

____

❷ What do you think these living things eat?

____

____

____

❸ Where do you think these living things sleep?

____

____

____

❹ Do you think you will see different plants and animals in different areas? Why or why not?

____

____

____

II. Observe It

❶ Go for a walk around your yard and your neighborhood. Take this *Laboratory Notebook* with you so you can draw and write your observations.

❷ Carefully observe the environment you're in by walking slowly and looking up to the sky, down to the ground, and everywhere in between. Look at things close up and from farther away.

❸ Look for different animals, bugs, birds, plants, and people. Observe what they are doing. Are birds flying in the sky, singing in trees, or hopping in the grass? Is a dog taking its person for a walk? Are there flowers blooming? Is a squirrel chattering at you from a tree? Are there bugs crawling on plants? Do you see any animals, birds, or bugs eating? What are they eating?

❹ Use the following boxes to draw and write what you see. There is a space in the gray boxes where you can note where you are when you make your observations.

Observations of the Environment

Observations of the Environment

Observations of the Environment

III. What Did You Discover?

❶ Did you observe anything about the plants in your environment that you hadn't noticed before? What did you observe?

____

____

____

❷ What did you see animals doing?

____

____

____

❸ Were any of the living things eating? What and how were they eating?

____

____

____

❹ When you walked from one area of your environment to another, what differences in plants, animals, and birds did you observe?

____

____

____

❺ What did you see that was surprising?

____

____

IV. Why?

The biosphere contains all the living things on Earth. Within the biosphere are different areas called *environments*. Each environment has a particular set of resources that make it possible for certain living things to live in that area. For example, a desert environment is a good home for lizards, cacti, and other animals and plants that don't need much water. An ocean environment is where you will find plants, fish, and animals, such as whales and dolphins, that can live in salty water. There are pine forest environments, mountain environments, lake and river environments, coastal environments, plains environments, and many others.

The region where you live has its own environment. The weather, amount of rainfall, type of soil, variations in temperature, plants, animals, and insects all work together to create this environment.

An environment can be any size. As you were making observations on your walk, you may have noticed that certain plants grow best in one area but not another, and certain animals live in one area but not another. This is a result of differences within the larger environment. Each plant and animal grows best when it finds just the right conditions to live in.

V. Just For Fun

An exoplanet is a planet that is orbiting a different sun (star) than ours.

Imagine you are the first astronaut to land on the exoplanet Kepler-62e in the constellation Lyra. Suppose you discover that Kepler-62e has a biosphere.

What do you think you would see when you land? Do you think Kepler-62e would look like Earth or be very different? Would there be an atmosphere with clouds? What would they look like? Would you find water? Where would the water be found? Would there be lots of plants and animals? What would the plants look like? What would the animals look like? What would they be doing? What would they eat? Would there be human-like creatures? What would they look like? What would they be doing? What else do you think you would see on Kepler-62e?

On the next page draw or write what you imagine the biosphere on Kepler-62e would look like.

A Visit to Kepler-62e

Experiment 16

Moving Iron

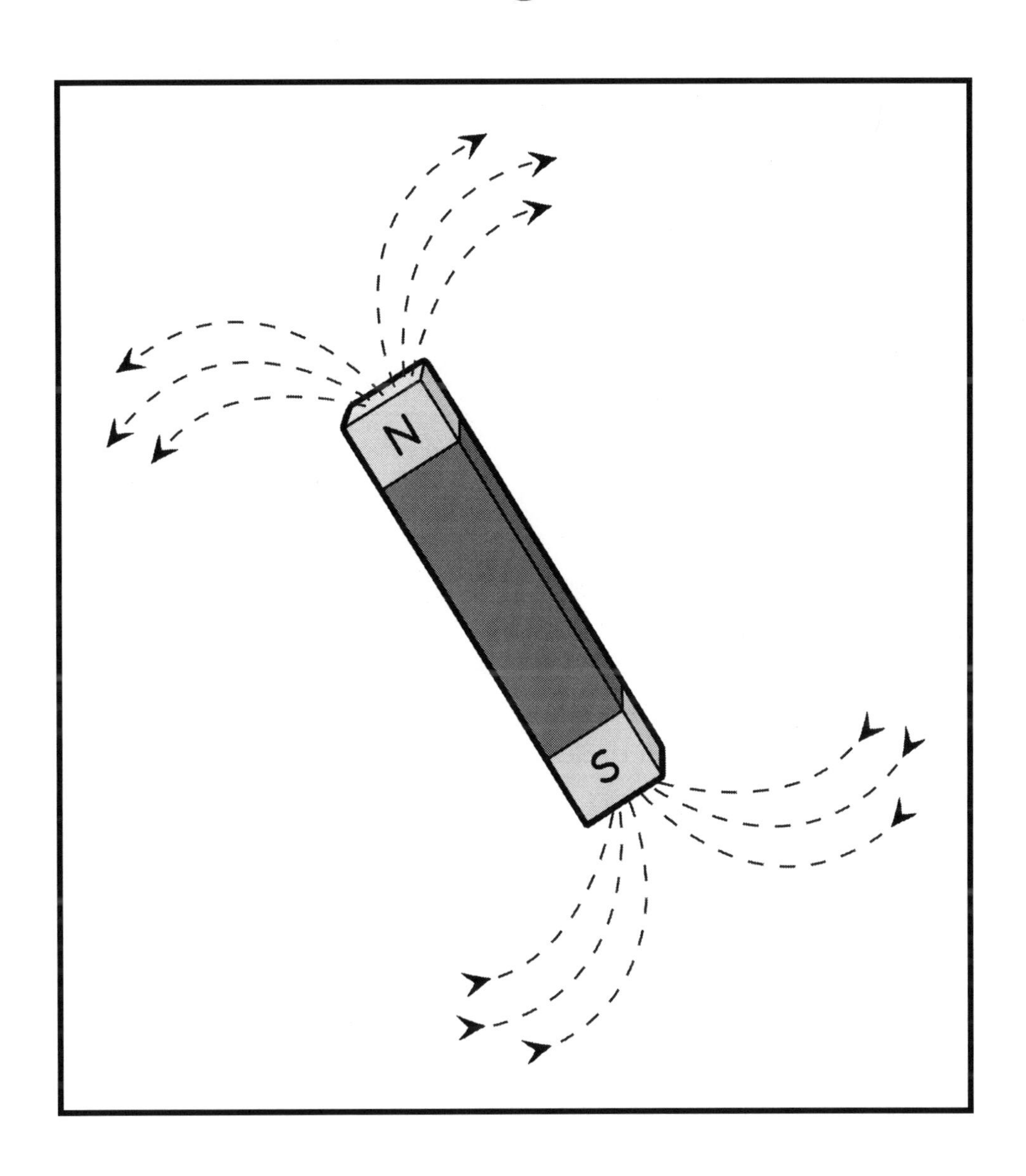

Introduction

Do you think there is a way to see the magnetic field created by a magnet? Find out with this experiment.

I. Think About It

1. Do you think if you look at a magnet you will see its magnetic forces? Why or why not?

____

____

____

2. Do you think magnetic forces can be useful? Why or why not?

____

____

____

3. If you have two magnets, do you think you could make the two north poles stick together? Why or why not?

____

____

____

4. Do you think there is a way to show that Earth has a magnetic field surrounding it? Why or why not?

____

____

____

II. Observe It

❶ Pour corn syrup into a plastic box until there is about 6 millimeters (1/4 inch) of syrup covering the bottom.

❷ Put the bar magnet on the table and place the box on top of it so the magnet is about in the center of the box.

❸ Pour iron filings on top of the syrup, being careful to not breathe them in.

❹ Wait 30 minutes and then check the iron filings.

❺ In the box below, draw what the iron filings look like.

III. What Did You Discover?

❶ What did the iron filings look like when you first put them in the syrup?

____

____

____

❷ What did the iron filings look like after 30 minutes?

____

____

____

❸ What do you think made the iron filings move?

____

____

____

❹ Do you think if you did this experiment again, the iron filings would end up in the same pattern? Why or why not?

____

____

____

❺ What do you think this experiment might tell you about Earth's magnetic field?

____

____

____

IV. Why?

A magnetic field is the area around a magnet that is affected by magnetic forces. In this experiment you were able to "see" the magnetic field around your magnet by observing how it affected the iron filings.

Earth has a magnetic field similar to the one that affected the iron filings in this experiment. Earth's magnetic field points out from the North Pole, surrounds the Earth, and points in at the South Pole. Earth's magnetic field is different from that of a bar magnet because it is thought to be created by swirling molten metals in Earth's core rather than from the atoms in a solid bar of metal.

Another difference is that Earth's magnetic field extends into space and is affected by energy from the Sun. When the Sun's energy hits the magnetic field, it creates the magnetosphere. The magnetosphere, in turn, lets enough heat and light energy through to keep plants and animals healthy. At the same time, it stops too much energy from getting to Earth. Life could not exist if too much of the Sun's energy reached the surface of Earth, so the magnetosphere is essential for life to exist.

V. Just For Fun

❶ Try moving the box so the magnet is in a different position under it. Wait about 30 minutes, then look at the pattern the iron filings make. Did repositioning the magnet make a difference in the pattern?

In the box below draw the pattern made by the iron filings.

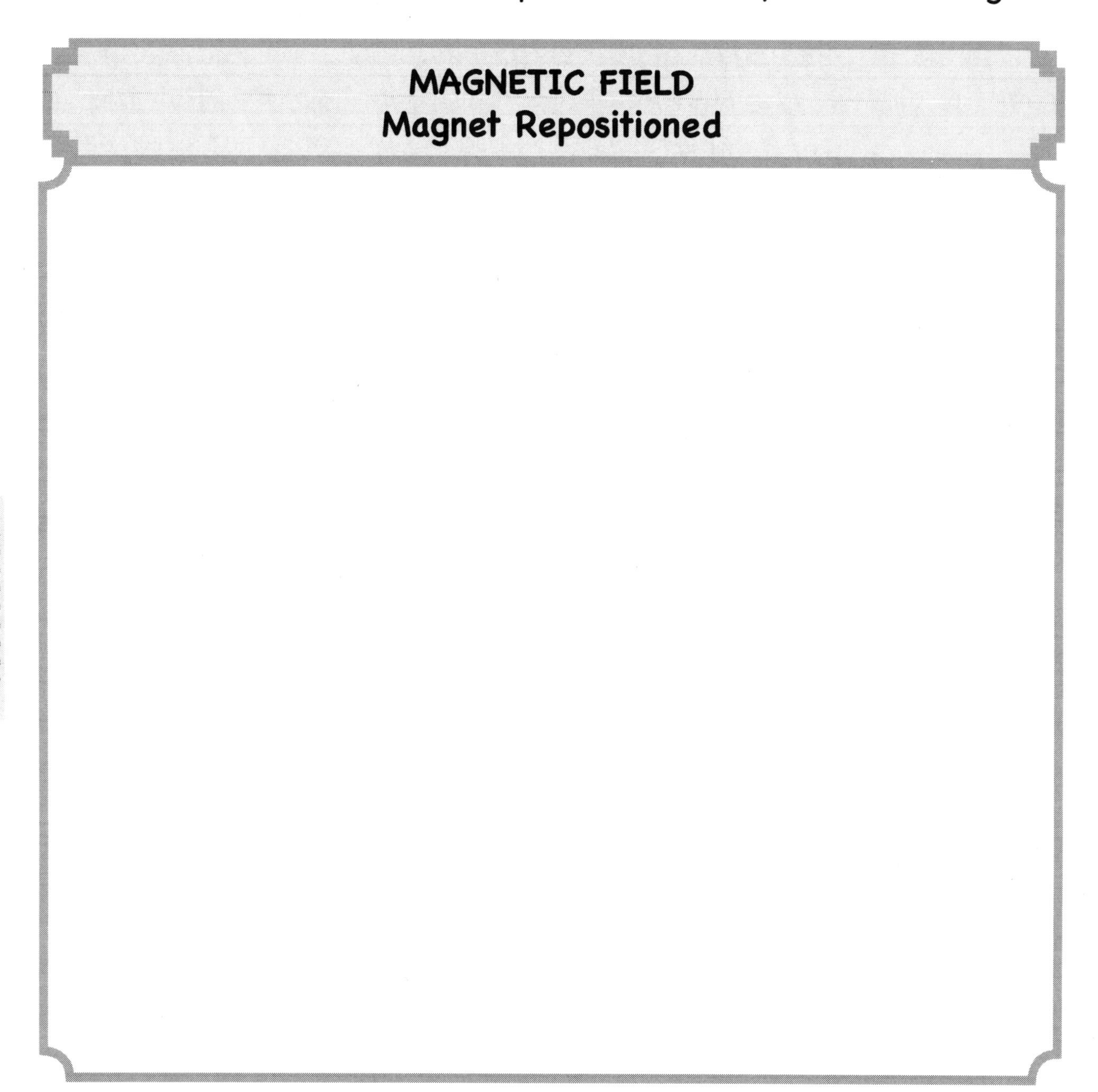

❷ Place a second magnet under the box. Wait about 30 minutes. Does having two magnets change the pattern of the magnetic field?

Draw your observations below.

❸ Try moving the two magnets to different positions. Wait about 30 minutes each time, then check to see what the pattern of the iron filings looks like.

Draw your observations below.

Experiment 17

What Do You Need?

Introduction

Do you think you can observe how the different spheres of Earth work together? Try this experiment.

I. Think About It

❶ Do you think a plant needs the atmosphere in order to live and grow? Why or why not?

____

____

❷ Do you think a plant needs the hydrosphere in order to live and grow? Why or why not?

____

____

❸ Do you think a plant needs the biosphere in order to live and grow? Why or why not?

____

____

❹ Do you think a plant needs the geosphere in order to live and grow? Why or why not?

____

____

II. Observe It

Plant a garden!

❶ Decide what kind of plant you want to grow and then get seeds for it.

❷ Plant several seeds outdoors in a garden. If you don't have space for a garden, you can use a flower pot, milk carton, or other container with soil in it.

❸ Water the seeds when you plant them and then check the soil daily. When it starts to get dry, add more water.

❹ Write or draw your observations in the following boxes. Each box can have observations for more than one day.

❺ As your plant gets bigger, what do you observe about its growth? How fast is it growing? Is it staying healthy? Is it getting too much or too little water or sunshine? Are bugs eating the plant? If so, what do they look like? Do you see any worms in the soil? Are any animals or birds affecting your plant? What else can you observe?

❻ Write or draw your observations as the plant grows. As you make observations about your plant, think about which of Earth's parts is affecting it and in what way. Is it the biosphere, hydrosphere, atmosphere, geosphere, or magnetosphere, or is it a combination of two or more of these?

GROWING A PLANT

GROWING A PLANT

GROWING A PLANT

GROWING A PLANT

III. What Did You Discover?

❶ How easy or difficult was it to grow your plant? Why?

❷ Did your plant get enough water from rain? How could you tell?

❸ Did your plant have any problems with bugs? Why or why not?

❹ Were any animals helpful or harmful to your plant? Why or why not?

5. Was there enough sunlight for your plant? How can you tell?

____

____

____

6. Was the soil you used good for your plant to grow in? Why or why not?

____

____

____

7. In what ways do you think the different parts of the Earth worked together to help your plant grow?

____

____

____

____

____

____

____

____

IV. Why?

For a plant to grow, all the different parts of the Earth have to work together. The plant is part of the biosphere and so are the bugs, animals, and people that might eat the plant. Bacteria in the soil fix nitrogen for plants to use, and worms add nutrients to the soil.

Earth's geosphere provides soil that has minerals the plant needs for making food. The soil contains water that is taken up by the plant's roots. The soil also provides a place for roots to anchor the plant to the ground so it won't blow away in the wind.

The hydrosphere provides the water needed by the plant, whether it comes from rain or from your garden hose. The hydrosphere works with the atmosphere to make clouds that move over the Earth, carrying rain to different parts of the land.

The atmosphere has carbon dioxide for the plant to use to make food. The gases in the atmosphere hold onto some of the heat energy from the Sun, helping to keep Earth warm at night. Also, the atmosphere lets light energy from the Sun reach Earth so plants can use the Sun's energy for making food.

The magnetosphere protects the plant from getting too much energy from the Sun, which would be harmful.

If any one of these parts of Earth were missing, plants could not live and grow.

V. Just For Fun

Grow an herb garden!

Herbs are plants that are great for adding extra flavor to salads, soups, and other foods. Sometimes they are used to make delicious teas and for medicines.

Herbs can be grown outdoors in pots or in a garden bed, and many can be grown in small pots indoors.

Decide which herbs you would like to grow. You might look for information about herbs in the library or online. Or you might want to choose them by names that sound interesting or by what the plants in the store look like. Some possibilities are: basil, rosemary, thyme, chervil, lemon balm, oregano, parsley, dill, spearmint, chamomile. If you have a cat, it might enjoy having some catnip.

Once you have decided which herbs you want to grow, get seeds or small plants.

Tend your herb garden, and when the plants are big enough, gather some leaves to put in your salad and soup. While you and your family are eating the tasty herbs, you can tell everyone how all the parts of the Earth worked together to bring each plant's unique flavor to you—Earth's geosphere, biosphere, hydrosphere, atmosphere, and magnetosphere all helped out!

Experiment 18

Modeling a Galaxy

Introduction

Building a model is a great way to learn more about galaxies.

I. Think About It

❶ How many neighborhoods do you have in your city or a city that you have visited?

❷ What else is in the city other than neighborhoods?

❸ What is at the center of a city?

❹ How is a galaxy like a city?

____

____

____

____

❺ What do you think is at the center of a galaxy?

____

____

____

____

❻ What do you think is at the center of the universe?

____

____

____

____

II. Observe It

1. Making a model involves first thinking about the different features you want to represent and then deciding which materials you will use to represent these features.

   Think about the different objects that make up a galaxy. List below the objects you want to represent and the materials you will use to represent them. Do you think all suns and planets look the same as each other?

Objects	**Materials**

❷ On a clear, flat surface place the materials you chose for modeling a galaxy.

❸ Design your galaxy. Where do the solar system "neighborhoods" go? How far apart are they? What else is in your galaxy? What is in the center? Write and/or draw your ideas below.

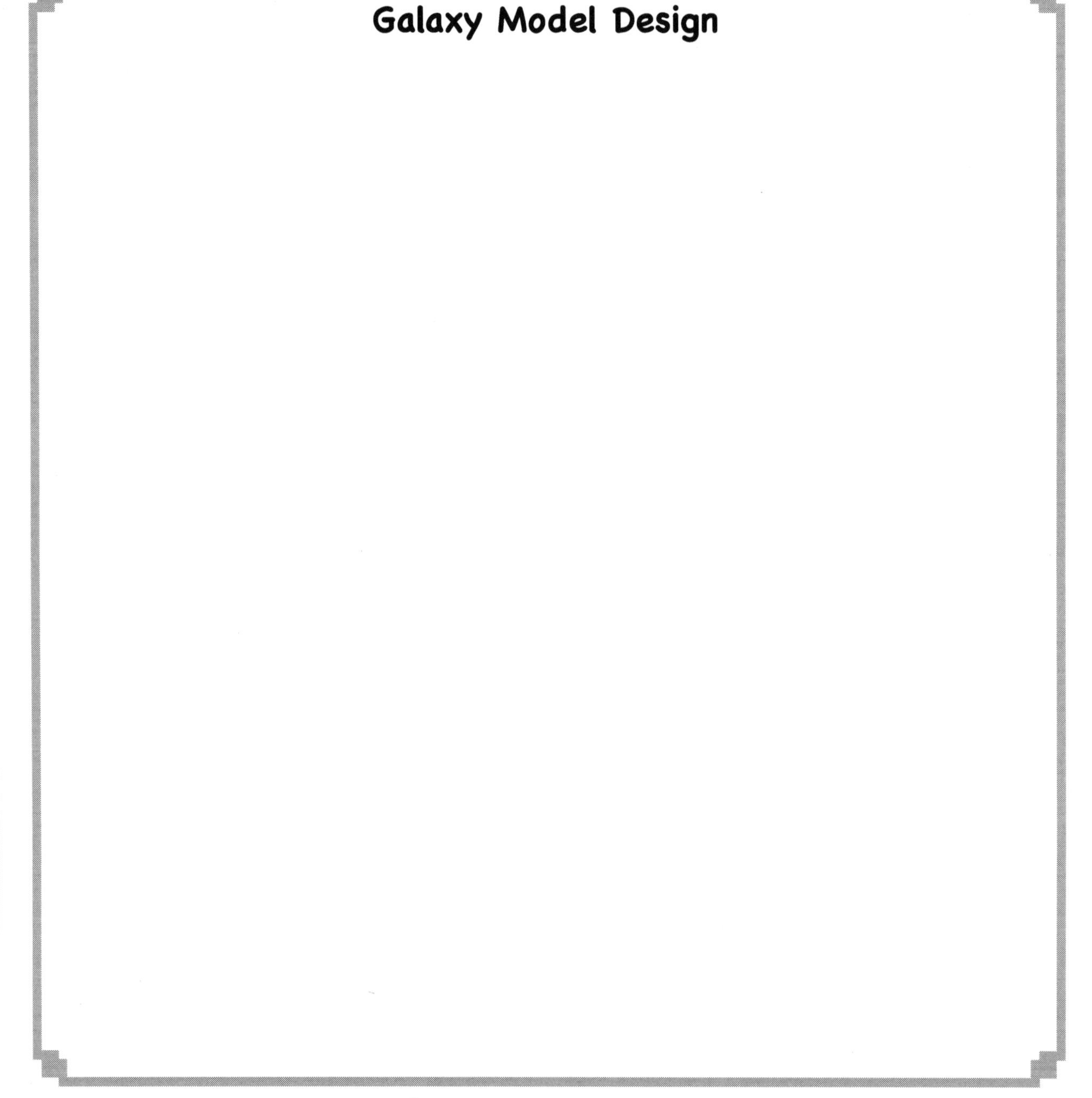

❹ On a piece of cardboard or poster board, create a galaxy as you designed it.

❺ In the box below record any unique features of your galaxy.

Galaxy — Unique Features

❻ In this box you can tape a photograph you have taken of your galaxy model, draw it, or write about it.

III. What Did You Discover?

❶ How many solar systems were you able to put in your galaxy?

____

❷ How did you decide where to put the solar systems and other objects?

____

____

____

❸ Did you run out of room? Why or why not?

____

____

____

❹ If the universe holds more than 170 billion galaxies, how big do you think it is?

____

____

____

❺ Do you think the universe will ever run out of room? Why or why not?

____

____

____

IV. Why?

Building models is a great way to help us think about things that have different features and many parts. In this experiment you built a galaxy and modeled it. You probably discovered that your galaxy was limited by the size of your cardboard or poster board. The size of the materials you chose to represent the various parts of the galaxy also limited how many of these parts you could include in your model.

We know that cities on Earth run out of room because mountains, rivers, oceans, deserts, and other features of the landscape restrict the amount of space available. Sometimes a city runs out of room because it bumps into the city next to it. This brings up an interesting question that scientists don't know the answer to—does the universe run out of room? Does it have an end? Are there another universes next to it?

V. Just For Fun

Imagine you could travel to the end of the universe. What do you think it might look like? Do you want to include some of the things you saw while getting there?

Draw and/or write your ideas in the following box.

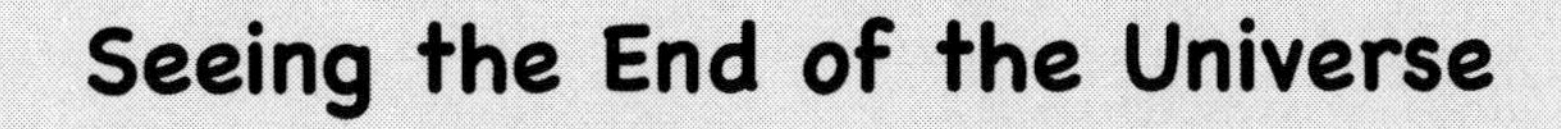
Seeing the End of the Universe

Experiment 19

See the Milky Way

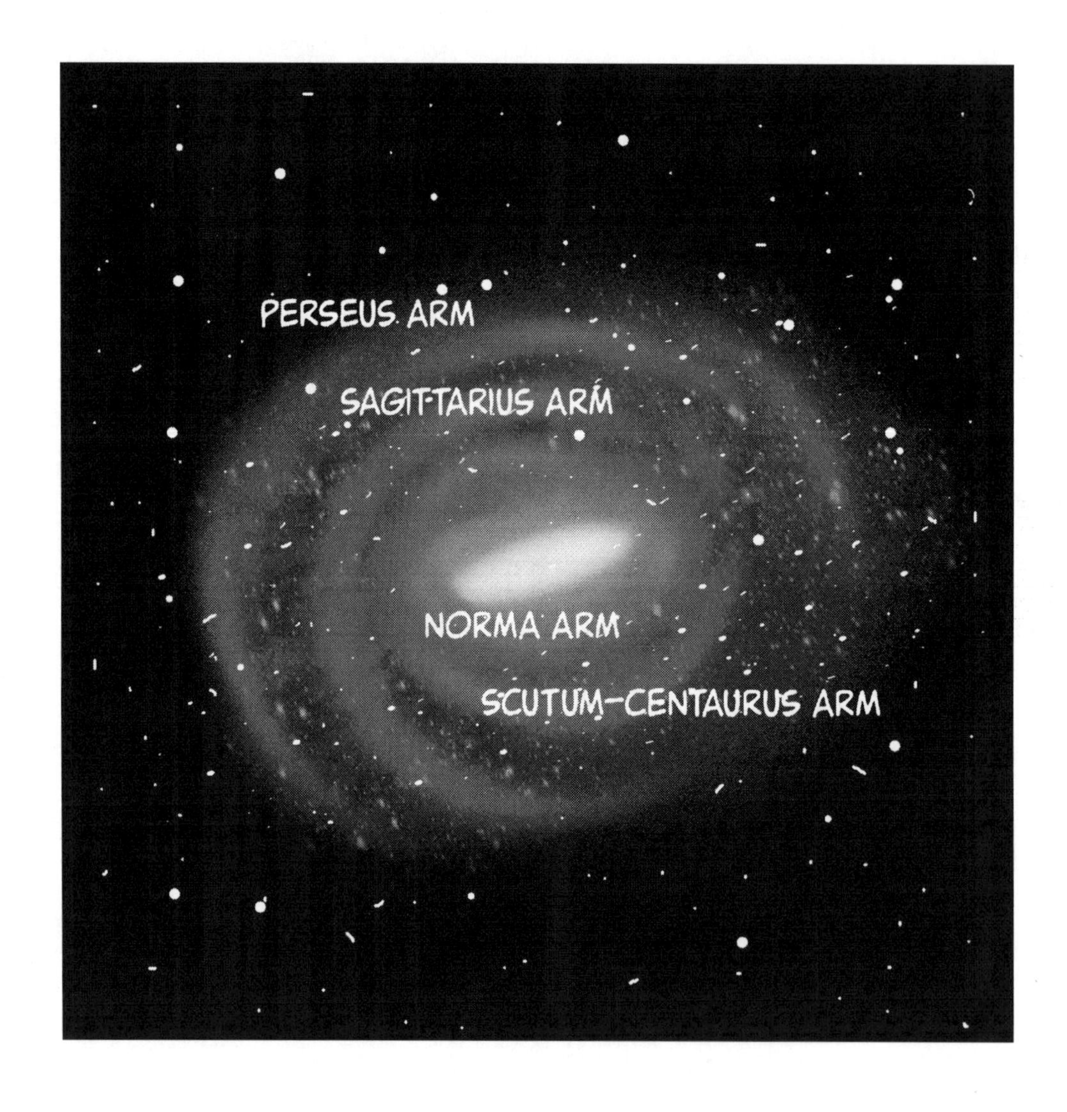

Introduction

In this experiment you will look at the Milky Way.

I. Think About It

❶ Looking at the Milky Way is like looking at your city. If you are near the edge of your city and you can see tall buildings from where you are, look for the area that has the most tall buildings. This will probably be the center of your city.

If you can see the center of your city, draw it below. If you can't actually see the center, draw what you think it might look like. If you don't live in a city, think about one you have visited or seen in movies or pictures and draw what you think a city center looks like.

❷ When you look away from the center of the city you will probably see fewer buildings. If you can look away from the center of your city, draw what you see below. If you can't actually see away from the center or if you don't live in a city, draw what you think the part of the city that is farther away from the center might look like.

❸ Let the buildings in the center of your city represent stars in our galaxy. In the space below, draw what the stars in the center of our galaxy might look like.

❹ Letting the buildings that are farther away from the city center represent stars, draw what you think the stars that are farther away from the center of our galaxy might look like.

II. Observe It

1. Find an area that is free of city lights on a clear night when there is no moon.
2. Look into the night sky without a telescope or binoculars. Use only your eyes.
3. Study the sky and observe areas where there are lots of stars. Compare this to areas with fewer stars.
4. See if you can find a band of light and stars stretching across the night sky.
5. In the space below, draw what you observe.

III. What Did You Discover?

1. How many stars did you see?

____

____

2. Were there areas with lots of stars and other areas with fewer stars? How would you describe the areas of the sky?

____

____

____

3. Were you able to see a band of stars stretching across the night sky? If so, how would you describe it?

____

____

____

4. If you could see this band of stars, do you think you were seeing the center of the Milky Way Galaxy, the edge of the Milky Way Galaxy, or something else? Why?

____

____

____

____

____

IV. Why?

Earth has an atmosphere that allows us to look through the air to see the stars, Moon, planets, and other objects in our Milky Way Galaxy. Most of the stars and other objects we see in the night sky are part of our galaxy. Earth is located at a perfect spot within the Milky Way Galaxy to observe what surrounds us.

When you see a narrow band of light and stars stretching across the night sky, this band is referred to as the Milky Way. To see this, you are actually looking through the spiral arms of the Milky Way Galaxy toward the center of the galaxy. Because the Milky Way Galaxy is a flat, disk-shaped spiral and we are looking at it edge-on, the stars you observe as you look through the spiral arms appear as a band of light across the sky. If we lived closer to the center of the galaxy, we would see so many stars all around us that it would be difficult to know which way to look to see toward the center.

V. Just For Fun

If you have a computer and would like to see the Milky Way Galaxy, download Google Earth from the internet. Follow the setup instructions. Click on the planet symbol at the top, choose "Sky" from the drop down menu, and type "Milky Way" in the search box.

What did you discover? On the next page write or draw what you found out.

Milky Way Discoveries

Experiment 20

How Do Galaxies Get Their Shape?

Introduction

Galaxies are groups of stars, planets, comets, asteroids, dust, and other things, such as gases. Scientists think that the force of gravity causes all of these objects in space to clump together in groups that have different shapes. Gravity is the force that holds everything together in a galaxy.

I. Think About It

❶ How do you think galaxies form?

❷ How do you think planets and stars are held together?

❸ What do you think causes spiral galaxies?

____

____

____

____

____

____

❹ What do you think causes irregular galaxies?

____

____

____

____

____

____

II. Observe It

In this experiment you will simulate the force of gravity on stars and other objects in space by using a magnet to move small metallic particles. Magnetic force is different from gravitational force but similar enough to use it to model galaxy formation.

❶ Take a shallow, flat-bottomed plastic container and fill it to just below the top with corn syrup.

❷ Pour the iron filings on top of the syrup, being careful to not breathe them in.

❸ Carefully cover the container with plastic wrap.

❹ Place two magnets underneath the plastic container and observe the iron filings. Record your observations in the space below.

❺ Take one of the magnets and create a swirling pattern. Record your observations in the space below.

❻ Take both magnets and create opposite swirling patterns Record your observations below.

❼ Bring the two magnets together and observe what happens. Record your observations.

8. Play with the magnets and iron filings. Try moving the magnets in different ways. Record your observations.

Magnet Movement: ______________________________

____

Magnet Movement: ______________________________

____

ASTRONOMY

Magnet Movement: ______________________________

____

III. What Did You Discover?

❶ What happened to the iron filings when you placed the magnets below them?

❷ When you swirled one magnet, did spiral arms form? Was there a center?

❸ What happened when you brought the two magnets together and allowed the iron filings to follow?

❹ Were you able to create any irregular shapes? Describe below what you did.

IV. Why?

Galaxies form because the gravitational forces of stars pull on each other and on planets, comets, asteroids, ice, and dirt. When a force pulls on an object, the object will begin to move.

In this experiment you built a model using iron filings and magnets to observe what is possible when forces move objects. You were able to see how magnetic forces pull on iron filings to create different shapes. In much the same way, the gravitational forces of stars pull on each other and on planets and other objects in space to create the shapes of galaxies.

V. Just For Fun

Make a Jell-O galaxy.

With the help of an adult, follow the instructions on a box of flavored gelatin. Add grapes, berries, or other fruits cut into small pieces. Before the gelatin cools, swirl the fruit into a spiral galaxy, bar galaxy, or irregular galaxy. How many different kinds of galaxies can you make?

In the following box, record your observations

Jell-O Galaxies

Experiment 21

Making a Comet

Introduction

A comet is a mixture of dirt and ice. When a comet travels close enough to the Sun, the ice will vaporize, turning into gas. This gas then creates a tail that follows the comet as it moves through space.

I. Think About It

❶ Draw what you think a comet in space might look like when it is far away from the Sun.

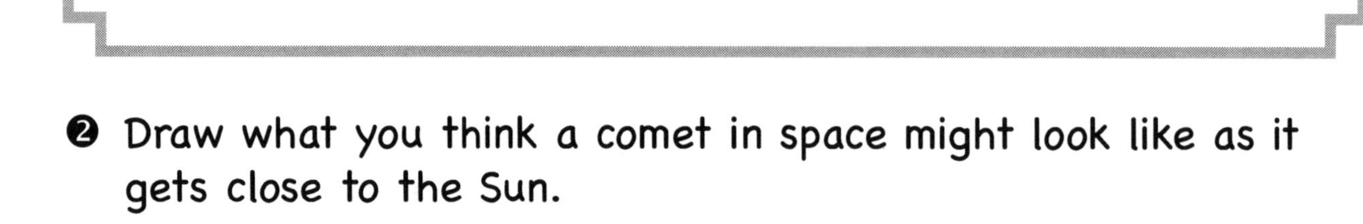

❷ Draw what you think a comet in space might look like as it gets close to the Sun.

II. Observe It

❶ Collect some dirt and small stones.

❷ Pour the dirt and stones into a small pail and cover them with water. Leave several inches between the water and the top of the pail.

❸ Place the pail in the freezer and allow the water to freeze.

❹ Tap the frozen model comet out of the pail.

❺ Observe the frozen model comet. In the space below, draw or write what you see.

❻ Observe the model comet as it melts. Draw or write your observations.

❼ Repeat Steps ❶-❹ using more water. Record your observations.

❽ Repeat Steps ❶-❹ using more dirt. Record your observations.

III. What Did You Discover?

❶ Do you think your frozen mixture of water, dirt, and rocks looks like a real comet? Why or why not?

❷ What happened as your comet melted? Did it come apart in chunks, or did it melt slowly?

❸ How quickly do you think your comet would come apart if it were near the Sun? Why?

❹ How much ice do you think a comet would need to have for you to be able to see its tail from Earth? Why?

____

____

____

____

____

❺ How much bigger than your comet model do you think a real comet is? Why?

____

____

____

____

____

IV. Why?

In this experiment you modeled a comet. Comets are large chunks of ice and rock that move through space. A real comet might look similar to the small comet model you made from ice, dirt, and rocks, but it would be much larger.

The ice in your comet model melted, but in a real comet the ice would vaporize, turning into gas without becoming a liquid first. Although the method by which the comet loses its ice is different in your experiment than it is for a real comet, this model lets you see what happens to a comet as it loses ice. It gets smaller and begins to break apart until the comet eventually disappears.

Scientists are not always able to make models that work exactly like the object they are modeling. But by making substitutions, scientists can still make valuable observations about objects they cannot get close to.

V. Just For Fun

With the help of an adult, make a water, dirt, and rock mixture and then add dry ice to it. How does the dry ice change your comet?

Experiment 22

All Science

Introduction

Most modern technologies are possible because scientists from different disciplines share their information. How do you think knowledge was combined to create computers?

I. Think About It

Think about the computer.

❶ What is the keyboard made of? What does it do?

____

____

____

____

❷ What is the screen made of? What does it do?

____

____

____

____

❸ What gives the device power? What different ways can you think of to power a computer?

____

____

____

____

❹ How are computers used in medicine?

❺ How are computers used in space?

❻ How do computers help geologists?

II. Observe It

Over the next week, observe how computers are used every day in games, toys, machines, cars, and other devices. Record your observations below and on the next page.

Computer Observations

More Computer Observations

III. What Did You Discover?

❶ Are computers used in games? If so, how?

❷ Are computers used in toys? If so, how?

❸ Are computers used in the kitchen? If so, how?

❹ Are computers used in cars? If so, how?

❺ Are computers used on bicycles? If so, how?

____

____

____

____

❻ Are computers used to grow food? If so, how?

____

____

____

____

❼ Are computers used in medicine? If so, how?

____

____

____

____

IV. Why?

Today, computers are used in almost every walk of life. The computer and the internet are possible because of mathematics, physics, chemistry, and computer science. Computers are used in every area of science and medicine. Each scientific discipline develops new programs and uses for computers. Today, computers help scientists make new discoveries, and as scientists discover more and look for answers to new questions, computers are developed that are better, faster, and smaller.

V. Just For Fun

Even the smallest toys sometimes contain a computer chip. Using the library or internet, look up what a computer chip is and why it can be used in so many different items. Record what you discover. There's more room on the next page.

Microchip Discoveries

____

____

____

____

____

____

____

____

More Microchip Discoveries

More REAL SCIENCE-4-KIDS Books by Rebecca W. Keller, PhD

Focus Series unit study program — each title has a Student Textbook with accompanying Laboratory Workbook, Teacher's Manual, Study Folder, Quizzes, and Recorded Lectures

Focus On Elementary Chemistry
Focus On Elementary Biology
Focus On Elementary Physics
Focus On Elementary Geology
Focus On Elementary Astronomy

Focus On Middle School Chemistry
Focus On Middle School Biology
Focus On Middle School Physics
Focus On Middle School Geology
Focus On Middle School Astronomy

Focus On High School Chemistry

Building Blocks Series year-long study program — each Student Textbook has accompanying Laboratory Notebook, Teacher's Manual, Lesson Plan, and Quizzes

Exploring the Building Blocks of Science Book K (Coloring Book)
Exploring the Building Blocks of Science Book 1
Exploring the Building Blocks of Science Book 2
Exploring the Building Blocks of Science Book 3
Exploring the Building Blocks of Science Book 4
Exploring the Building Blocks of Science Book 5
Exploring the Building Blocks of Science Book 6
Exploring the Building Blocks of Science Book 7
Exploring the Building Blocks of Science Book 8

Super Simple Science Experiments Series

21 Super Simple Chemistry Experiments
21 Super Simple Biology Experiments
21 Super Simple Physics Experiments
21 Super Simple Geology Experiments
21 Super Simple Astronomy Experiments

Kogs-4-Kids Series interdisciplinary workbooks that connect science to other areas of study

Physics Connects to Language
Biology Connects to Language
Chemistry Connects to Language
Geology Connects to Language
Astronomy Connects to Language

Note: A few titles may still be in production.